The
OCEANS
are
EMPTYING

The
OCEANS
are
EMPTYING

Fish Wars and Sustainability

Raymond A. Rogers

BLACK ROSE BOOKS

Montréal/New York
London

Black Rose Books No.
Hardcover ISBN: 1-551640-31-7
Paperback ISBN: 1-551640-30-9
Library of Congress No. 95-79350

Canadian Cataloguing in Publication Data

Rogers, Raymond Albert
The oceans are emptying: fish wars and sustainability

Includes bibliographical references and index.
ISBN 1-551640-31-7 (bound). —
ISBN 1-551640-30-9 (pbk.)

1. Sustainable fisheries. 2. Fishery policy. I. Title.

SH328.R64 1995 333.95'6 C95-900716-4

Cover photo by Chris Van Dyck

Mailing Address

BLACK ROSE BOOKS
C.P. 1258
Succ. Place du Parc
Montréal, Québec
H2W 2R3 Canada

BLACK ROSE BOOKS
340 Nagel Drive
Cheektowaga, New York
14225 USA

A publication of the Institute of Policy Alternatives of Montréal (IPAM)

Printed in Canada

for
Laura Jane McLauchlan

Acknowledgements

I want to thank my colleagues in the Faculty of Environmental Studies at York University for their helpful comments and encouragement, especially my former advisor Paul Wilkinson, John Livingston, Rodger Schwass, Doug MacDonald, David Bell, as well as Anita McBride, for her good-spirited creation of academic community in the faculty. I also want to express my gratitude to my colleagues in that other school, the North Atlantic Ocean, who I fished alongside of from Little Harbour, Shelburne County, Nova Scotia: Dale Richardson, Bill Williams, Willy Harding, Joe Richman, Arthur Swansburg, David Stewart, Merton Swansburg, Chester Freeman, Frank Decker, Murray Swim, Milton Smith, Andrew Ringer, Robert Ringer, Cyril Giffin, Bud Hardy, Wesley Swansburg, Jeff Swansburg, Steve Swansburg, Terry Morash, Clyde Murphy, Borden Williams, Carman Williams, and John Richardson.

I wish to thank everyone at Black Rose Books: Dimitri Roussopoulos, Linda Barton, Natalie Klym, Frances Slingerland, and Robert Kwak, for their contributions in publishing this book. I would also like to thank the Sasakawa Peace Foundation for financial support during the writing of this book.

CONTENTS

I

Sustainability and the Dis-integration of Conservation and Development in the Northwest Atlantic Fishery

This inquiry began in the mid-1980s when the hooks started coming up empty. Nights that I spent in my boat, twenty miles off the south coast of Nova Scotia, setting two miles of long-line gear with 3,500 baited hooks on it were followed by days of hauling the gear back with a catch of less than 500 pounds of fish. Only a few years before, catches of 1,500 to 2,000 pounds were a regular occurrence. Although this inquiry considers a wide range of issues about the relationship between human society and nature, I always return to the question of why the hooks come up empty. It is a fisherman's question. It is a question which comes from the margins of modern society, and is intended to illuminate and contextualize the practices of modern society. Finally, it is also the question asked by someone who has been watching the Atlantic coastal community in which he lives decimated due to the depletion of the fish. In the context of modernity in general, it is the dead ends, the blind alleys and the collapses, which perhaps can evoke the destructiveness of a modern economy's standard practices.

Proceeding from the fisherman's question, this book rests on the premise that the sustainability debate related to the conservation of the natural world is an exceedingly recent, and largely theoretical discussion as it applies to modern human exploitation levels. This starting point is linked to a second proposition which states that the goals and strategies of conservation in the literature pertaining to the management of the Northwest Atlantic fisheries, and the goals and strategies of the literature on global sustainability, are very similar. When this close similarity in theoretical approaches to the conservation of nature is combined with the profound fail-ure of conservation practices in the Northwest Atlantic (as well as in other ocean fisheries), there arises the possibility that this failure in practice in the conservation of the ocean fishery can be a valuable basis for assessing the goals and strategies of sustainability generally. With regard to the relationship between theory and practice then, the North Atlantic fishery represents a useful case study about the failure of the practice of conservation, which in turn can generate useful discussions about the theory of sustainability.

It is this destruction of the fish species, and the accompanying vul-nerability of the human communities that depend on them, which underlies the "Turbot War" that took place in the spring of 1995 between Canada and Spain. The high-seas confrontations just outside Canada's 200-mile limit reflect the conflictual relationship between conservation and development that appears in the aftermath of ecologic collapse. There is every indication that "environmental war" is becoming an increasingly common response to ecologic failure, not only in the oceans, but in a variety of different milieus. Ecologic breakdown therefore becomes a precondition to social and political breakdown. Environmental problems are becoming increasingly complex not just because they interact in biophysical terms, but also because they create instability in the larger societal context. Once we enter the world of ecologic collapse, there is no easy road back to social and ecologic viability. It is there-fore incumbent upon us to address environmental problems before we enter this unpredictable and strife-ridden world, and not after.

Since the Canadian government declared the moratorium on fishing for northern cod on the Grand Banks (July 1992) due to the ecologic col-lapse of the northern cod — which has been called a "great destruction" of biblical proportions for the people of Atlantic Canada — attention has focused on seal predation, changing climatic conditions, and foreign over-fishing as possible causes for the economic and ecologic crisis in the fishery, as well as some begrudging admission of regulatory failure. Within the fish-ing industry itself, the inshore blames the offshore, the long-liners blame the draggers and gillnetters, the small companies blame the multi-nationals, and everyone blames the Canadian government for mismanaging the industry. As much as there might be an element of truth to each one of these claims, the overwhelming evidence is that — despite the range of approaches used to regulate the industry — ecologic collapse was caused by the fact that too many fish went to market. Or, to put it another way, standard practice in a modern economy destroyed one of the most fertile fishing grounds in the world. If we are to deal with environmental problems before ecologic collapse, rather than after, I believe it is necessary to find a way to discuss environmental problems that makes the structures and processes of modern everyday life problematic, and does not accept them as givens.

This book will therefore focus on the relationship between standard practice in a modern economy — or what is called development in sustainability literature — and the various strategies that were used in the fishery to integrate development with conservation initiatives. The history of comprehensive attempts to regulate the exploitation of nature in the name of conservation is brief and dismal. I believe however, that the significant aspects of the current sustainability debate are all present in the history of regulation in the Northwest Atlantic Fishery. First and foremost, there is the full knowledge, recognized by all involved, that unregulated exploitation leads to ecological collapse. In the early 1970s, the international distant-water fleet made up of the industrialized countries of the world exploited the fish in the Northwest Atlantic to the point of collapse. There is no question here that regulators and exploiters were unaware of the possible consequences of human action. In fact, all the justifications for regulation — especially the declaration of the 200-mile limit — arise out of the recognition of this initial calamity.

Attempts at regulation began nominally in 1949 with the creation of the International Commission for the Northwest Atlantic Fishery (ICNAF) which accompanied the expansion of the international distant-water fleet after World War II. This increasingly powerful industrial fleet exploited a wide range of fisheries across the globe in the 1950s and 1960s. In the aftermath of the ecologic collapse of the fish in a number of fisheries, including the Northwest Atlantic in the early 1970s, various attempts were made in the international context to enforce gear and quota restrictions. But, because

ICNAF was a voluntary organization and had no enforcement capability, these attempts to control exploitation in the mid-1970s failed.

The failure to regulate this increasingly powerful international fleet led directly to the world's coastal states declaring 200-mile Exclusive Economic Zones in 1977. This declaration was based on the claim that international law was insufficient to insure the conservation of marine ecology, and that the depletion of marine communities could only be overcome by internalizing regulatory approaches within the circumscribed boundaries of the nation state. International open-access patterns were to be replaced by the "unified directing power" of the national mandate which would provide both economic stability for domestic fleets and ecologic stability for marine biotic communities. As they were later expressed in the sustainability debate, this national mandate was explicitly trying to integrate the perspectives of conservation and development from the very beginning.

Throughout the period of national management in the Northwest Atlantic from 1977 to 1992, there were periodic economic and ecologic crises in the industry. One of the most important crises led to the Kirby Task Force in 1982, the outcome of which was the amalgamation of a series of financially-troubled fish companies — companies which had expanded rapidly after 1977 and were caught with large inventories and no cash flow when the recession of the early 1980s hit — into large vertically-integrated companies such as Fisheries Products International in Newfoundland, and National Sea Products in Nova Scotia. Through a massive infusion of capital, the Canadian government set about creating a "modern" industry that would overcome the backwardness and the poverty of the past.

The priorities of large industry began to dominate fishery policy in official terms. This led to comprehensive regulatory frameworks being gradually replaced by the privatization of fish quota, beginning with the larger players in the industry. By the late 1980s, it became clear that — with great effort and at great expense to the Canadian taxpayer — the Canadian Department of Fisheries and Oceans was using a range of conservation strategies to manage the fishery right out of existence.

The single most important factor in understanding the regulatory failure in the Northwest Atlantic fishery is that development proceeded to the point of ecological collapse. Only in the aftermath of collapse did the initiatives for conservation begin to be put in place. This observation has important ramifications for my argument about addressing environmental problems before ecologic collapse rather than after, and the connection this has with dominant assumptions of modern economic development.

As defined within resource management perspectives, the central role of conservation is two-fold: to develop a biological information base in order to assess and understand the natural communities that are being exploited, and to regulate exploitation levels on an ongoing basis so that they remain within the limits of those biological realities. When development occurs first, the opposite becomes the case. Since there is little or no biological information that precedes the collapse, any analysis after the collapse takes place in the context of dis-equilibrium where the biotic community is even more difficult to understand; secondly, newly-created management frameworks have to cope with the fact that the forces of development which caused the collapse are already in place, and are resistant to initiatives that lower the levels of exploitation.

The poor quality of biological information and the continued strife between regulators and exploiters, which have been ongoing problems in Canada's East Coast fishery, have their root in this situation where development precedes conservation. What adds an element of tragedy to all this is that — by declaring the 200-mile limit and thereby expelling the international fleet — Canada had the opportunity to bring development under control. Instead of doing so, the Canadian government promoted the processes of destruction by funding the expansion of the Canadian fleet. This repeated in the national context what had previously collapsed marine communities in the international context.

It is my intention to illustrate that these realities in the fishery have a great deal to contribute to the sustainability debate. First and foremost, I believe we have to recognize that, in global terms, we are living in the aftermath of collapse. There is every indication that development — as Wolfgang

Sachs states — is an "intellectual ruin" from which we have to extricate ourselves.[1] In the *State of the World 1995*, Lester Brown asserts that when "sustainable yield thresholds are crossed, the traditional responses proposed by economists no longer work."[2] In the world of unsustainability, the instruments that are set forth to lead modern society towards sustainability, such as the internalizing of externalities, pollution permits, effluent taxes, voluntary regulation — and it is this debate which dominates current sustainability discussions — are eclipsed by social and political realities. In the world of unsustainability, the instruments of conservation are replaced by the gun-barrel and high-seas confrontations.

After taking part in the negotiations which led up to the United Nations Conference on Environment and Development, the *Earth Summit* in Rio de Janeiro (1992), Pratap Chatterjee and Matthais Finger concluded that "as a result of the whole UNCED process, the planet was going to be worse off, not better" and that "the outcome is a new push for more environmentally destructive industrial development."[3] Most of the analysis of the problems in the fishery occurring in the aftermath of collapse, reflects a similar failure to come to terms with the destructive aspects of industrial development. I regard the events that have occurred in the Northwest Atlantic fishery leading to its ecologic collapse as typical of the way in which the forces of development precede and then overrun conservation initiatives. In presenting the dis-integration of conservation and development in the Northwest Atlantic fishery, it is my intention to challenge sustainability practitioners and policy-makers with this question: What is it that you would do differently so that the world will not end up in the same calamity that now confronts the Atlantic coastal communities — caught between ecological collapse and the other, more fickle source of life and death, the Canadian government?

It is not that I am trying to point the finger in such a way that excludes myself or any other individuals or groups from taking responsibility for the destruction of marine biotic communities. I have taken part in this destruction. Although my role in this whole process was a small one — I fished for twelve years in a 35-foot inshore long-line and lobster boat — I certainly understand the motivations and pressures to fill the fish hold on a daily basis. And in this regard, I am reminded of Bubby Thorburne, a Shelburne fisherman who was well-known for his ability to catch fish. Bubby and his crew drowned south of the Grand Banks in a storm in December of 1989. A northwest gale came up quickly and caught them. This was not the first boat that Bubby had lost, although it was the first time that people died.

Thorburne had a reputation for steaming back from the Grand Banks with the boat swamp-loaded with fish and only six inches of free-board showing above water. He sank one boat on a fine day while hauling gear. The trip was just about over, and the crew were hauling the last string of long-line gear. The holds were full below decks, and above decks the fish were mounded up and sliding back over the rails into the water. The crew slogged around hip deep in fish. They had even filled the bunks, as well as the life boat, which they were towing just off the stern of the boat. Bubby's brother was still hauling the gear at the rail, shacking off all but the largest steak cod when the boat went down. As a society, this seems to be our plight: we will carry on doing what we normally do until the boat goes down.

DEVELOPMENT:
MAKING A DEAD THING GROW

Unfortunately the basic economic principles underlying the exploitation of renewable resources are still very poorly understood by decision makers. . . . Perhaps the most important initial realization for the question of sustainable development is that the overwhelming environmental and resource problems now facing humanity are the result of *economically rational individual decisions* made everyday by each and every one of us.

Colin Clark[4]

The seeming intractability of the forces of development, as well as the inability of regulatory frameworks to bring the forces of exploitation under control, provides the central focus of this inquiry. There are two aspects to this inability to control exploitation. First, there is the recognition that members of modern industrial society are legally destroying the very fabric of planet Earth. Second, even when regulation is instituted to control exploitation, it is circumvented in one form or another, so that the end result is ultimately the same. This is a profoundly depressing basis upon which to discuss the relationship between modern human society and the natural world. However, this calamitous view of human-nature relations is derived from my personal experiences and my understanding of the failure of regulation in Canada's East Coast fishery. It is my position then, that this failure in the fishery is entirely typical of the processes associated with modern economic and technological realities. The various conservation strategies used in the fishery are also entirely typical of those associated with resource management and sustainability. Accompanying these failures in conservation practices, is the inability to understand the reasons for these failures.

Most of the goals of regulation are based on the myth that public regulatory bodies operate in terms of equity and fairness as they oversee the private economic competition for resources, goods, and services engaged in by citizens and corporations. The dividing line between public and private is a porous membrane through which the edicts of economic development pass with impunity. This is especially true when it comes to other beings on this Earth (besides humans), who lack the voice and legal standing to have an impact on human demands. This public-private basis for discussions on regulation is inadequate for understanding modern environmental problems. The reality is that government regulation exists for the most part to promote development, and almost all legal structures are predicated on assumptions about the freedom of the individual, private property, and the expansion of capital. It is these realities which contribute to the intractability of environmental problems, since these realities are already in place when environmental problems occur, and resist being circumscribed by conservation initiatives.

A more viable approach to environmental problems comes out of discussions on the subject of hegemony, and the recognition that the structures of everyday life and the structures that cause environmental problems are one and the same. Hegemony refers to the sum total of beliefs, institutions and lived practice of a society which, although very complex, presents a dominant perspective which is far deeper than what is generally considered an ideological assumption. Raymond Williams describes it in this way:

> This notion of hegemony as deeply saturating the consciousness of a society seems to me to be fundamental. And hegemony has the advantage over general notions of totality [of social reality], that it at the same time emphasizes the facts of domination.[5]

In assessing "the facts of domination" with regard to the modern exploitation of the natural world, Colin Clark's contention that "resource management is a special problem in capital theory" is a very useful starting point. Clark states:

> Recognizing the capital-theoretic nature of resource stocks is essential to a clear understanding of resource economics. From this viewpoint resource management simply becomes a special problem in capital theory, although it is an especially interesting and difficult problem.[6]

While from the point of view of economics and development, exploitation of nature is a *special problem* in capital theory, from the point of view of the conservation of nature, capital theory is *the problem*. In the context of late capitalism and the intensification of the structures and processes of globalization, capital theory is not only putting excessive pressure on the natural world, it also permeates the very way we understand who we are as humans and how we relate to nature. In the essay "Ideas of Nature," Raymond Williams describes the social and economic realities associated with capital in terms of being a massive "interference" in human relations and human-nature relations.[7]

In Canada, the largest government department is now the Federal Department of Human Resources. Resources for what, I ask? The only answer is, resources that serve the expansion of capital. If we think of ourselves in this way, how can we possibly think of nature in anything but resourcist terms? And as John Livingston states, once any part of nature is deemed useful to the community of capital, "its depletion is only a matter of time."[8] It is this kind of perception which can be linked to the realities of hegemony.

There are two sides to this reality. First, there is the recognition that the way we understand the relationships between humans, as well as the relationships between humans and the rest of nature, are circumscribed by modern economic realities. Second, these economic realities promote the destruction of human and natural communities. In other words, we are at risk of losing the ability to talk about what is destroying us. At the heart of modern society is a dead thing that grows (capital). Living things die to make a dead thing grow. But then again, capital is not a thing, but a set of relationships in which we participate. As such, we grant it a reality and a power over us. As we participate in these relationships, we are remade by them and become the resources that serve the growth of a dead thing.[9]

In the Northwest Atlantic fishery, this attempt to regulate the exploitation of marine biotic communities has led to management at the end of a gun barrel. This is because conservation measures have been unable to stop the expansionist logic of capital theory. To quote Pitcher and Hart,

> It should be emphasized that the MPV [maximization of present value] takes virtually no account of the biological properties of the stock, except in as far as predicted yields are used to calculate values. . . . No one can claim that a policy which deprives their grandchildren of the fishery is a rational one. Unfortunately, in a free economy, the MPV

process seems to happen virtually automatically without it necessarily being a conscious aim of management.[10]

Although many have argued that the competition for common property resources has led to the pressure to maximize present values due to the fact that no one can be assured that the fish will be there tomorrow as they may be caught by someone else, Clark claims that economic decisions under an open-access framework and those made under private property can yield the same results:

> We have presented the basic theory of the dissipation of economic rent in the unregulated open access fishery, and we have compared that theoretical situation with the harvest policy of a profit-maximizing sole owner.
>
> The principal conclusion of our analysis is that as a consequence of time discounting, the sole owner also "dissipates" a certain fraction of the potential maximum economic rent by achieving an optimal level of sustained rent that is less than the maximum.
>
> The extent to which a resource owner elects to "dissipate" sustainable rent (in favour of immediate revenues) depends on the rate of discount employed in the calculation of present values.[11]

Clark links this pressure on natural communities to the idea that they are standing capital, subject to the pressure of being converted into a commodity so that it may return as more capital:

> The sole owner considers a resource asset to be a form of capital. If the asset fails to provide a suitable return on its capital value (in comparison with alternative forms of investment), it is profitable to liquidate the asset. If whaling companies expect to earn 10% on their investments and if the blue-

whale population only increases 5% per annum, the whalers could quite rationally plan to liquidate the blue-whale stocks.[12]

Robert Heilbroner presents this process as part of the defining characteristic of capital relations:

> This repetitive, expansive process is, to be sure, directed at bringing goods and services into being through the organization of trade and production. But the physical attributes of these commodities, even when they take the form of luxurious objects, are not prized as evidences of a successful completion of the search for wealth, as long as they are in the capitalist's possession. On the contrary, their physical existence is an obstacle that must be overcome by converting the commodities back into money. Even then, when they are sold, the cash in turn is not regarded as the end product of the search but only as a stage in its never-ending cycle [of expansion].[13]

It is precisely this expansive process which regulatory agencies and those who wish to manage development can do nothing about. The essence of capital is expansion: "Without the organizing purpose of expansion, capital dissolves into material building blocks that are necessary but not sufficient to define its life purpose."[14] With regard to the exploitation of the natural world, it is this "life purpose" conveyed by Heilbroner which exerts hegemonic domination over both the realities of exploitation and most of the discussions concerning what to do about it. It is also this "life purpose," or should I say death purpose, which by its very definition is not amenable to limitations in any way. We are currently seeing this purpose expressed in the edicts of privatization, deregulation, and free trade which dominate current discussions on global relations. The absorption of human identity within these realities encloses us further within the forms of exploitation which are becoming increasingly pervasive in late capitalism. As Marx states:

> . . . [I]t is only in so far as the appropriation of ever more and more wealth in the abstract becomes the sole motive of his operations, that he functions as a capitalist, that is, as capital personified and endowed with consciousness and a will.[15]

In *Nature and the Crisis of Modernity*,[16] I have described this process in terms of a double disappearance: not only is the natural world disappearing in real terms, but the human impetus to mount arguments in favour of the conservation of nature because it is part of "us" is also disappearing. Almost all conservation measures initiated over the last two decades have been ineffective because they are conceived and operate within the structures of development, and thus do not make modern economy sufficiently problematic. In other words, conservation requires us to challenge the hegemonic structures of the modern world.

It is the refusal to face these realities, and the assumption that they can be regulated, that has tarred sustainable development with the "business as usual" stick. As long as a modern, consumer-based and growth-oriented perspective is brought to bear on the Earth, there will be species depletion, habitat destruction, and pollution, no matter what the management framework. Within the hegemonic structures of late modernity, there can be no equal and opposite force for conservation to counter the demands of development.

This perspective is confirmed in an analysis of fishing activity in the Northwest Atlantic and its relation to the management frameworks that have been put in place to regulate it. Why have the fish gone from being seemingly infinite in numbers to being in a state of collapse over a period of thirty years, with the added quandary that this plunder was achieved by an industry that stumbled from crisis to crisis? During that time, politicians and managers have struggled to make the fishery more industrialized and modernized, so that it would operate like other industries in the market economy. The Kirby Task Force objectives state:

> The Atlantic fishing industry should be economically viable
> on an ongoing basis, where to be viable implies an ability to
> survive downturns with only a normal business failure rate
> and without government assistance.[17]

Instead of trying to find ways to make the fishing industry operate like other businesses, as was the goal of the Kirby Task Force, it is my intention to question "normal business" and its relation to the destruction of the natural world.

CONSERVATION:
STRATEGIZING IN THE AFTERMATH OF COLLAPSE

> In biological conservation we cannot hope to "get something
> for nothing." If future yields are to be increased, current har-
> vest levels must be reduced. The fundamental problem then
> becomes one of determining the optimal trade-off to be made
> between current and future harvests. This problem, which is
> the very essence of resource conservation, is an exceedingly
> difficult one, not from a mathematical viewpoint, perhaps,
> but certainly from a political and a philosophical viewpoint.
>
> Colin Clark[18]

As opposed to resource management, which approaches conservation in the context of the creation of goods and services for human consumption, or sustainability — which defines itself in terms of meeting present human needs without limiting the possibility of future generations meeting theirs — this book is based on the recognition that, if the ecologic viability of other species on this Earth are to be preserved, a conception of conservation which is merely an aspect of the production process is not enough to insure this protection. What is required is a conception of conservation which makes the edicts of development sufficiently problematic so as to challenge them directly in terms of their most basic assumptions. Alternately, what is also required is a socially-viable sense of both the human community and the natural community which forms the basis of this challenge to development, and its accompanying conceptions of humans and nature as the material which serves the expansion of capital.

But what is meant by the word conservation, anyway? In terms of the "wise use" perspective, which in no way challenges the modern economic paradigm, conservation is almost synonymous with efficiency, albeit over a longer time frame. Because conservation is conceived of in terms of efficiency, there is no reason why the market should not look after it, since being efficient is what the market does so well. From other less functional perspectives, conservation is conversation, it is a way of talking about development. To speak of the "social" is to find a way to talk about the "economic." To talk of "diversity" is a way of conveying the increasing "monoculture" of technological society. Finally, to talk about "local" is to struggle to find a way to behold "global" realities. These are perspectives on problems, not solutions. They are local, socially-viable, and diverse, conservation-oriented ways of grappling with economic, monocultural, and global forms of development. In other words, conservation creates an analytical space in which to talk about development. The historical forces that are causing environmental problems have been a long time in the making, and are all but identical with the standard practice of modern everyday life. To talk about conservation of the natural world is therefore, by definition, to begin to make the structures and processes of modern everyday life problematic. The extent to which these structures and processes are made problematic defines the range of the environmental debate. To talk about the relationships in a functioning multispecies community in which "all participants are subjects"[19] is to raise the corollary in modern human society that, in late capitalism, increasingly "all participants are objects" that serve the growth of a dead thing.

Arthur McEvoy describes the kinds of obstacles to conservation initiatives in the California fishery which he associates with the structures and processes of modern life:

Throughout most of its history, U.S. law worked in service to the private economy to dissolve whatever barriers either the ecology of the resources themselves or the efforts of some fishers to stabilize their relations with the fish might place in the way of sustained expansion. The fundamental autonomy and irresponsibility of market actors, which lay at the core of the fisherman's problem, was an article of constitutional faith, the perceived foundation of liberty and opportunity for harvesters. When they did attempt to curtail fishery use, lawmakers had to depend on the political favor of the industry for whatever power they had to influence the course of events. The fisherman's problem thus reproduced itself in the very structure of policy processes which were supposed to correct it.[20]

Adding to this statement, McEvoy comments further ". . . the fisherman's problem consists as much of people stealing from each other as it does of people stealing collectively from nature."[21] It is this set of relations which can be directly attributed to modern market relations, or what Karl Polanyi referred to as disembedded relations.[22] Furthermore, this insight highlights the close connection between conservation and the broader social and political context.

So when we look for solutions to environmental problems, can we solve isolated problems, or do we have to solve history? Because there is a sense of immediacy about most environmental problems, there is a tendency to jump to solution mode without due regard to the causes of problems. It does not necessarily follow from the recognition that modern human lifestyles are unsustainable, that there are sustainable modern life styles. To merely flip the negative components of a problem into positives does not necessarily yield solutions. The sustainability debate is about unsustainability, not about sustainability, and therefore requires, I believe, more definitions of problems and less ascendant, future-oriented solutions. And we do a great disservice to the severity of these problems by not recognizing how deeply entwined they are with modern human "reality."

If globalization is the latest stage in modern development, the goals of privatization, deregulation, and free trade reflect the increasing triumph of capital. Because the human project is becoming increasingly dominant in the world, and represents itself almost entirely within resourcist categories, it becomes all but impossible to think of nature or humans in any other terms but those associated with the edicts of development. This enclosure of conceptions of humans and nature within the realities of late capitalism locates the environmental crisis within a wider crisis of modernity. To talk of conservation without addressing this crisis of culture is to abandon it to the forces which are destroying the natural world. In this context, conservation based on a radical inter-species egalitarianism, as expressed in environmental thought and philosophy for example, is a significant perspective because of its sharp contrast to other initiatives which so manifestly serve the modern human project. It is also significant because it implicitly challenges so much of what is taken for granted in modern life.

This contrast between privileging the human project and inter-species egalitarianism becomes even more important when we recognize the fact that most conservation initiatives which have defined themselves in terms of development have at most times failed to insure economic or ecologic viability. So, in discussing the failure in the fishery, I use it as an example of how conservation goals, as they have been defined in terms of the current resource economic paradigm, have great difficulty in protecting the natural world — and the links that failure has with conservation biology, resource management and sustainability generally — so as to challenge many of the current perspectives on the relationship between conservation and development.

Resource management was initially concerned with the efficient conversion of natural resources into goods and services. In response to the appearance of problems related to pollution and depletion, resource management became a more complicated decision-making process. Not only were there new kinds of ecological information about natural communities which had to be considered, but there also appeared more complicated multi-stakeholder decision-making fora which began to include those who had suffered from the externalities of the resource management process.

As it has evolved within resource management frameworks, conservation has been concerned with a process that values the non-economic aspects of the conversion process. Narrow, short-term economic considerations have been broadened to include the long-term detrimental costs of the pollutants which are dispersed into the surroundings. The costs of replenishing exploited natural resources has also been considered. This ensures that that the producer bares the "real" long-term cost involved in the production of goods and services. What this has meant in terms of resource management and sustainability is a recognition that methods are required which give value to externalities such as pollution or depletion so that they can be included as variables within the economic model. Once again, I believe that this kind of conservation strategy does not make development sufficiently problematic because it converts human values as well as those associated with nature, into categories which must ultimately suit the modern economic model — in effect, universalizing the realities of capital by swallowing accompanying conceptions of humans and nature.

All of the sustainability literature and resource management literature assumes that if you have a sophisticated enough mathematical model and adequate biological information, as well as the economic model which includes these values, it is possible to set sustainable exploitation levels. This perspective assumes that if you know nature well enough, a dotted line will appear in nature which says, here is the principle and here is the interest, so do not exploit beyond this line. By its very definition, sustainability assumes this level exists, and that society can operate at this cliff edge of exploitation. However, that level is no where in nature. It is an entirely human, entirely economic, number. Nature is not a factory which produces an annual surplus for exploitation which can be skimmed off. This is a production model of nature which resides in modern humans and not in other beings. It is a profound social failure which conveys the extent to which nature has been objectified as the raw material of the production process.

This accommodation of development by conservation perspectives is conveyed in the definitions of conservation and development in the *World Conservation Strategy*:

> Development is defined here as: the modification of the biosphere and the application of human, financial, living and nonliving resources to satisfy human needs and improve the quality of life. . . .
>
> Conservation is defined here as: the management of human use of the biosphere so that it may yield the greatest sustainable benefit to present generations while maintaining it potential to meet the needs and aspirations of future generations.[23]

Because this conception of development is so benign — what could be more reasonable than improving the quality of human life? — the corresponding unenergetic conception of conservation need not be very exigent. *Caring for the Earth* — the 1991 update of the *World Conservation Strategy* — reflects a similar accommodation, while at the same time acknowledging that there are problems with certain kinds of development:

> We need development that is both people-centred, concentrating on improving the human condition, and conservation-based, maintaining the variety and productivity of nature. We need to stop talking about conservation and development as if they were in opposition, and recognize that they are essential parts of one indispensable process.[24]

These two statements about the relationship between conservation and development are entirely wrongheaded. Firstly, when *Caring for the Earth* states that "we need development that is both people centred. . . and conservation-based" what it is saying is that modern development is not people-centred nor conservation-based. As I discussed earlier, it doesn't follow that you can flip the negative of problems into the positives of solutions. If you examine why development is not people-centred or conservation-based, you would find deeply-rooted ideas and structures related to capitalism which the weak kneed definitions of conservation presented here can do nothing about. Contrary to the cries by *Caring for the Earth* to "stop talking about conservation and development as if they were in opposition," I contend that an oppositional view of conservation that resists development is precisely what is required.

Conservation therefore requires a far more antagonistic countermovement to the forces of development if it is going to resist the pressure of capital theory. Central to this project is the articulation of a social-ecological space in which conservation can exist. In other words, conservation requires a community in which to realize itself. This community has to have the capacity to resist the resourcist categories of modern economy, both for humans as well as for the rest of nature. Conservation can only begin in the implicit recognition of membership in a socially-viable community.

Once again, I contend that the failure of conservation in the Northwest Atlantic fishery confirms many of the shortcomings of conservation strategies discussed here. I will now discuss the context in which these conservation strategies were undertaken.

ECOLOGICAL BRINKMANSHIP

. . . the Canadian Government considers customary
international law inadequate to protect Canada's interest
in the protection of the marine environment and its
renewable resources.

Law of the Sea Conference[25]

The catching capacity of the international distant-water fleet which roamed the world's oceans increased dramatically after World War II as war ships were converted to draggers and technologies developed during the war such as powerful engines, hydraulic winches, radar, and sonar were put to use in the fishery. As the fish were depleted off the coasts of Europe, fleets of Russians, Poles, Spaniards, French, Portuguese, and German boats moved into the Northwest Atlantic off Canada's East Coast as well as to other continental shelves which had a good supply of fish. A volunteer umbrella organization called the International Commission for the Northwest Atlantic Fishery (ICNAF) was set up in 1949 to gather information on fishing activity in this specific area.

Throughout the expansion of this fleet in the 1950s and 1960s, ICNAF was generally understood to be working well as catches steadily increased. But in the aftermath of the collapse of marine communities in the late 1960s and early 1970s resulting from international exploitation, ICNAF failed at attempts to impose gear restrictions and country by country quotas during the mid-1970s. Because of the international context and the resulting lack of enforcement capability, ICNAF — as a volunteer, umbrella organization — failed to limit fishing effort and this led coastal states to unilaterally nationalize the 200-mile Exclusive Economic Zones in 1977 as a first step toward replacing "customary international law," while putting into place the regulatory framework needed to overcome the destructive exploitation patterns of the past. During the "Turbot War" between Spain and Canada in March of 1995, the shortcomings leveled at ICNAF previous to 1977, were now being directed towards the North Atlantic Fisheries Organization (NAFO) — which took over the monitoring of fishing activity in international waters outside the 200-mile limit — and its inability to limit either catch limits or gear use of the Spanish fleet.

This chain of events with regard to the Northwest Atlantic is reflected in the global fishery generally. In *World Fisheries Resources*, James R. Coull states:

> The process of intensification [of exploitation] has been particularly rapid since the Second World War, and the increased pressure on resources has resulted in frequent over-fishing;

this in turn was a major motivation behind the radical change in the International Law of the Sea, which replaced the former freedom of fishing with exclusive national limits extending out to 200 miles from the coast.[26]

Coull argues that the brinkmanship of exploitation patterns in the fishery makes it a very useful case study for regulatory responses to the pressure modern economic and technological realities exert on natural processes:

> . . . a sophisticated modern armoury of equipment and techniques. . . rendered more obvious the fragility of the resource base and the danger of damaging it by over-fishing. This has rendered resource conservation programmes necessary on a scale never seen before. . . . As the limit in yield has been increasingly closely approached in recent decades, a situation has developed in which fish have acquired a degree of importance beyond that which would be justified by their importance to fishing communities and fish consumers in different countries, and which has given them a notable measure of importance in national and international politics.[27]

In the case of Canada, this importance is reflected in the dual conundrum of the failure of domestic management of the fishery on the one hand, and international confrontation on the high seas on the other. It is the dire combination of species collapse and military conflict which points towards the increased occurrence of environmental war in the future in a range of social and political locales.

NATIONALIZATION OF MARINE LIFE AND THE EMERGENCE OF RESOURCE MANAGEMENT

The implementation of a resource management institutional capability formed the basis of the Canadian Government's and other coastal states' claims to the high seas 200 miles from their coasts. Within this national regulatory framework, resource management perspectives now became the context in which discussions of fisheries issues occurred. Unlike political economy critiques of capital and markets which struggle "to find a frame of reference to which the market itself is referable," to quote Karl Polanyi,[28] the critical ability of resource management perspectives is limited by its implicit acceptance of the workings of modern economy. Resource management is an approach which strategizes with — but does not question — the demands that appear on the market. As McEvoy states with regard to the limited mandate of resource management in the California fishery:

> . . . external to the theory [of resource management] were the forces that drove the harvest: demand, technology, and other variables were factors that fishery managers had to cope with, but were not variables to be controlled.[29]

It is precisely these forces of demand and technology which political economists identify as being central to understanding the predatory relationship between economic processes and natural processes.

To provide justification for the nationalizing of the high seas, the Canadian Government layed out its first comprehensive approach to the fishery on the eve of the declaration of the 200-mile limit in 1977. In the *Policy for Canada's Commercial Fisheries*, management goals were set out with the aim of overcoming the chronic economic and ecological instability which had plagued the fishery in the international context:

> - Obtain national control of the exploitation of fishery resources throughout a zone extending at least 200 nautical miles from Canada's coasts.
> - Institute a co-ordinated research and administrative capability to control fishery resource use on an ecological

basis and in accordance with the best interests (economic and social) of Canadian society.

- Develop a fully effective capability for the monitoring of information on resource and oceanic conditions, for the surveillance of fleet activity and for the enforcement of management regulations.[30]

It is worthwhile to compare Bruce Mitchell's normative model of the way resource management should take place with the above mission statement of the Canadian Government. Mitchell presents a resource management process whereby a "natural resource becomes a commodity or service as it is shaped by human attitudes, technology, financial and economic arrangements, and political realities." For Mitchell, this process should occur in three stages:

> 1) Resource Analysis- determines the quality, quantity, and availability, as well as demand for product.
>
> 2) Resource Planning- the actual decisions which allocate and set the conditions of resource development.
>
> 3) Resource Development- the process whereby the resource becomes a commodity or service.[31]

In these terms, first there is an analysis of the size and quality of the resource, then, once it is decided that the resource provides a viable basis for exploitation, a management framework is put in place to ensure economic and ecological stability. After these first two stages are complete, resource development begins and the resource is converted to a commodity or service available to society.

As well as presenting an "ideal" of the resource management process, Mitchell's model also provides an analytical tool for managers to assess the possible reasons for the success as well as the failure of particular instances of resource development. The management goals set forth in *Policy for Canada's Commercial Fishery* were in line with how resource management should take place as they relate to Mitchell's model. They are also in line with how earlier versions of sustainable development were presented, as in initiatives such as the *World Conservation Strategy*:

> -Determine the productive capacities of exploited species and ecosystems and ensure that utilization does not exceed those capabilities.
> -Adopt conservative management objectives for the utilization of species and ecosystems.
> -Ensure that access to a resource does not exceed the resource's capacity to sustain exploitation.
> -Maintain the habitats of resource species.[32]

I contend that the most important reality in the failure of the fishery — and this is all but universal in the exploitation of the natural world — is that the marine biotic community was "developed" to the point of collapse before there was any consideration that there should be resource analysis or resource planning. This situation of reversal is reflected in Mitchell's description of how the policy process happens in resource management:

> 1) Identification of a significant problem, for which either there is no policy or else present policies are inadequate.
> 2) Formulation of a policy which attempts to solve the problem.
> 3) Implementation of the policy.
> 4) Monitoring the effects of the policy[33]

Most times, the "identification of a significant problem" relates to overexploitation of a resource. What is clear in the case of the fishery, and resource management and sustainability literature generally, is that conservation initiatives instituted in the aftermath of collapse, and making use of the structures and processes of modern institutional and legal frameworks, failed to insure either economic or ecologic viability.

In fact, the very forces which destroy natural communities are beyond the mandate of conservation as practiced in sustainability and resource management literature. The reason development precedes conservation is that we have a difficult time acknowledging that the structures of everyday life and the structures that cause environmental problems are the same. Standard practice in a modern economy destroyed the fishery. In order to challenge this standard practice, conservation requires a more challenging and antagonistic conception if it is to resist the ongoing and ubiquitous incursions of development. This alternate conception of conservation requires a contrasting set of structures and processes of a different kind of everyday life if it is to be realized in practice.

OUTLINE OF CHAPTERS

The book is organized to reflect the idea that development preceded conservation in Canada's East Coast fishery, and that this reality is a significant one in the discourses of sustainability in general. It also represents an important obstacle in achieving either economic stability or preservation of other living beings on Earth. It also indicates the ongoing problems related to the fact that the exploitation levels are not considered problematic until after an ecological collapse. As a result, resource management and sustainability are conceived of and operate as ongoing crisis management, which most times, does not challenge the causes of environmental problems. So, by organizing the chapters in terms of the resource management model (albeit in reverse order), I am not expressing any great faith in it. More to the point, I am attempting to show that it failed to fulfill its own terms of reference in the fishery. Therefore, Chapter I discusses resource development related to the industrialization of the oceans, Chapter II analyses the various resource management frameworks which attempted to regulate exploitation in the fishery. Chapter III examines the stock assessment models used by fishery scientists to calculate the number of fish in the ocean. Chapter IV focuses on a discussion of the "problem" of common property in the fishery as a basis for analyzing the assumptions which inform resource management and modern economy generally. The Conclusion presents the political economy of depletion and dependence, and an approach that would begin to challenge increasingly global economic forces.

NOTES

1 Wolfgang Sachs. 1992. *The Development Dictionary*. London: Zed Books, p. 1-5.

2 Lester Brown. 1995. *The State of the World*. New York: Norton, p.15.

3 Pratrap Chatterjee & Matthais Finger. 1994. *The Earth Brokers*. New York: Routledge, p. 2.

4 Colin Clark. 1990. *Mathematical Bioeconomics*. New York: Wiley, p. v.

5 Raymond Williams. 1980. "Base and Superstructure in Marxist Cultural Theory." *Problems in Materialism and Culture*. New York Verso, p. 37.

6 Clark (1990:68).

7 Raymond Williams. 1980. "Ideas of Nature." *Problems in Materialism and Culture*. New York: Verso, pp. 67-85.

8 John Livingston. 1981. *The Fallacy of Wildlife Conservation*. Toronto: McClelland and Stewart, p. 43.

9 Raymond A. Rogers. 1994. *Nature and the Crisis of Modernity: A Critique of Contemporary Discourse on Managing the Earth*. Montreal: Black Rose. I have discussed the relationship between development and modernity extensively in *Nature and the Crisis of Modernity*, and presented what I considered to be the reasons for the intractability of exploitation. In fact, although *The Oceans are Emptying* was published later, most of its research on the fishery was done prior to the writing of *Nature and the Crisis of Modernity*. In sequential terms, *The Oceans are Emptying* is the case study out of which arose the general considerations about the relationship between modern economics and the natural world which are addressed in *Nature and the Crisis of Modernity*.

10 T. Pitcher & P. Hart. 1982. *Fisheries Ecology*. London: Croom Helm, p. 303.

11 Clark (1990:62).

12 Clark (1990:62).

13 Robert Heilbroner. 1985. *The Nature and Logic of Capitalism*. New York: Norton, p. 36.

14 Heilbroner (1985:37).

15 Karl Marx. 1959. *Capital*. Moscow: Foreign Language Publishing, p. 152.

16 Raymond A. Rogers (1994:2).

17 M. J. L. Kirby. 1983. *Navigating Troubled Waters: Report for the Task Force on the Atlantic Fisheries*. Ottawa: Minister of Supply and Services.

18 Clark (1990:31).

19 John Livingston. 1994. *Rogue Primate: An Exploration of Human Domestication*. Toronto: Key Porter Books, p. 111.

20   Arthur McEvoy. 1986. *The Fisherman's Problem: Ecology and Law in the California Fisheries, 1850-1980*. New York: Cambridge University Press, p. 253-254.

21   McEvoy. (1986:257).

22   Karl Polanyi. 1957. *The Great Transformation*. Boston: Beacon Press.

23   IUCN, UNEP, & WWF. 1980. *World Conservation Strategy*. Gland, (Section 1).

24   IUCN, UNEP, & WWF. 1991. *Caring for the Earth*. Gland, p. 8.

25   *Law of the Sea Discussion Paper*. 1974. Ottawa: Department of External Affairs, p. 3.

26   James R. Coull. 1993. *World Fisheries Resources*. London: Routledge, p. 41.

27   Coull. (1993:4).

28   Karl Polanyi. 1968. *Primitive, Archaic and Modern Economies*. [George Dalton ed.] New York: Doubleday Anchor, p. 174.

29   Arthur McEvoy (1987:295).

30   Fisheries and Marine Service. 1976. *Policy for Canada's Commercial Fisheries*. Ottawa: Department of the Environment. p. 63-64.

31   Bruce Mitchell. 1989. *Geography and Resource Analysis*. New York: Longman & Wiley, p. 3-5.

32   *World Conservation Strategy*. (1980:Section 7).

33   Mitchell (1989:6).

II

Resource Development:
The Industrialization of the Ocean Fisheries

. . . Resource development represents the actual exploitation
or use of a resource during the transformation of 'neutral
stuff' into a commodity or service to serve human needs
and aspirations.

Bruce Mitchell[1]

Establishing economic value requires the disvaluing of all
other forms of social existence.

Gustavo Esteva[2]

A s stated in the introduction, resource management theory and
practice outlines the stages of resource analysis, resource planning
and management, followed by resource development, as the three
processes through which a resource passes in becoming available to society.
In the case of the Northwest Atlantic, as well as virtually all the ocean fisheries, resource development preceded any systematic programs of resource
analysis or resource management. As a case in point for assessing resource
management and sustainability, the fish in the oceans pose interesting and
difficult problems, as well as reveal many of the issues associated with failures in conservation.

Until very recently, the fish of the high seas have been an unhusband-
ed, unowned supply of food existing outside national boundaries and beyond
any legal or institutional arrangement. The number of fish in the ocean was
unknown and the interrelationships of the biotic community were not

understood because the fish were all but invisible beneath an often inhospitable conjunction of natural forces. In fact, the fish in the ocean are only known and encountered, for the most part, in the context of human commercial activity.

Because of these conditions, the fishery offers a valuable illustration of the dynamics of development that in a more regulated situation would be less graphic. The twin fulcrums of technology and capital can be clearly seen as the forces that promoted the depletion of fish communities. In a world that is suffering from a pervasive depletion of natural communities, the events that have taken place in the Northwest Atlantic fishery — as well as ocean fisheries generally — may offer insights into the dynamics of this destruction.

Because of its promotion of dependence in coastal communities through its centralized regulatory infrastructure and its acceptance and promotion of the economic realities which have caused the depletion of biotic communities in the Northwest Atlantic fishery, the resource management perspective of the Canadian Government remains an impediment to significant analysis in the aftermath of ecologic collapse. By contrast, a problematic examination of the assumptions which inform the relationship between modern society and the natural world generally, and which are associated with development priorities, is an evocative basis for analyzing a range of failures, of which the ocean fishery is but one of many.

THE GLOBAL OCEAN FISHERY

Expanding upon a seemingly inexhaustible sea to provide protein for the hungry of the world, the industrialization of the global fishery in the thirty years that followed World War II led to a doubling of the catch of fresh fish to twenty million tonnes, an increase by a factor of ten in the catch of frozen fish (thirteen million tonnes) and a rapid expansion in the production of fish meal and oil (twenty million tonnes). A good deal of the exploitation by this international distant-water fleet took place in the Northwest Atlantic until there was ecologic collapse in the early 1970s. The distant-water fleets of Spain, Poland, Germany, Britain, Portugal, Russia, Japan, Canada, and the United States had built new stern trawlers that were equipped with innovations such as filleting machines, fish meal conversion units, freezers, hydraulic winches, and sonar fish finders. These powerful technologies very quickly decimated a range of ocean fisheries, including the Northwest Atlantic.

As well as onboard innovations that increased production, changes in the market place led to a higher demand for fish. As an infrastructure of freezers were put in place in the food distribution system, frozen fish products became available to a wider population. An increase in the demand for protein by a rapidly expanding human population was accompanied by an increase in the fish meal production for livestock feed necessary to serve the newly-created fast food sector. To quote Pitcher and Hart:

> The rapid increase in the quantity of fish caught was caused by the increasing world demand for protein, but this demand took two forms which stimulated two different types of fishery. As human populations expanded rapidly in the 1950s and 1960s there was an urgent need for more dietary protein at low cost. The capture of wild fish provided an attractive source that was economic and easy to obtain. . . .
>
> The second demand for protein came from increasingly industrialized pig and poultry farmers in the developed nations. . . . These industries used high-protein artificial diets and one of the chief sources of protein is fish meal.[3]

The demand for food by humans was more directly related to the demersal or bottom-feeder species such as cod and haddock which are caught predominantly by stern trawlers, while the demand for fish meal led to an increase in landings by purse seiners of pelagic fish such as anchovy, herring, and mackerel, which school near the water surface and feed on plankton.

These changes in technology and market demand made it possible for the global catch of fish to grow from 21.9 million tonnes shortly after World War II to 68.8 million tonnes in the early 1970s.[4] This period of open-ended expansion of fishing efforts based on industrialization ended in 1977, with the declaration of 200-mile economic zones by coastal states protecting the fish adjacent to their shores. Whereas earlier periods of industrialization based on such innovations as steam power remained localized events that affected relatively small areas, the industrialization that began in the 1950s made hitherto inaccessible fish open to exploitation and globalized the world fishery.

Fishing efforts continued to expand within national jurisdictions following the declaration of the 200-mile Exclusive Economic Zones, just as they had in the international open access context. Despite considerable expenditure by individual countries on fishery regulation, this domestic expansion threatened nearly every fishery in the world with ecological collapse. To quote Peter Weber in *State of the World 1995*:

> The catch has fallen in all but 2 of the world's 15 major marine fishing regions; in 4 of them, it has shrunk by more than 30 percent. . . . Analysts from the U. N. Food and Agricultural Organization (FAO) found overfishing in one third of the fisheries they reviewed; they found some depleted fish populations in nearly all coastal waters around the world.[5]

Despite the fact that the 200-mile limits were declared in the name of regulation and conservation, biotic communities continued to be overexploited. Weber attributes this overexploitation to practices encouraged by national economic development priorities. Weber states: "FAO estimates that countries provide on the order of $54 billion annually in subsidies to the fishing industry — encouraging its overexpansion in recent decades."[6] Coupled with this promotion of development is the fact that "government subsidies often favor large-scale operations"[7] which result in a situation where

"Modern equipment can provide the means, and commercial markets the motivation, for depleting fish stocks in ways that are not likely in traditional fisheries,"[8] and "if this continues, virtually the entire small-scale fishing sector could be wiped out."[9] Added to these national initiatives which lead to the overexploitation of the fish are the international agencies such as the World Bank which are "contributing to the overcapacity of commercial fisheries and undermining traditional fishers."[10]

In the Worldwatch Report entitled *Abandoned Seas: Reversing the Decline of the Oceans*, Peter Weber states that:

> overfishing has precipitated declines in individual stocks throughout the world. The catch of four commonly-eaten, average-value fish (Atlantic cod, Cape hake, haddock, and silver hake) fell from 5 million tons in 1970 to 2.6 million tons in 1989. . . . Fishers have managed to keep the marine fish catch climbing in past decades by abandoning fished-out stocks and pursuing new species. These substitutions, however, are typically lower-value fish that were previously undesirable and unwanted. . .[11]

This was precisely the case with the turbot fish which inhabits the edge of the continental shelf in the North Atlantic, and which initiated the "Turbot War" between Canada and Spain. This previously undesirable flatfish became more highly prized because it was one of the few fish left in the North Atlantic which was not in a state of ecologic collapse. Fighting over the crumbs is becoming increasingly common in many domains and forms the basis for environmental war. The oceans are at the point where there are fewer and fewer fish species, even of the previously undesirable type, to maintain current catch levels.

In many countries around the world — such as Sierra Leone in West Africa — overfishing occurs because national governments do not have the resources to patrol their coastal waters sufficiently in order to monitor intrusions by foreign vessels. In other areas — such as the state of Kerala in India

— the central national government engages in joint ventures with foreign fleets so as to produce much needed exports, while the artisanal fishers who supply the local markets of Kerala have their livelihood threatened due to overfishing by these foreign fleets.

What is particularly striking about the state of the ocean fisheries around the world is that despite the range of regulatory responses to exploitation, either within the context of national management strategies or in the international arena, the results have been all but uniformly bleak. This range of regulatory responses to exploitation is also represented in the specific history of the failures of conservation in the Northwest Atlantic. As James Coull states in *World Fisheries Resources*:

> While management measures and systems have developed
> and proliferated in the last quarter century, in the broad view
> the situation of global fisheries resources continues to give
> cause for serious concern. Situations of over-fishing have
> often persisted and multiplied.[12]

This spectre of overexploitation becomes especially daunting when we consider that Coull goes on to state that the Atlantic Canadian fishery has one of the "best systems of monitoring and management in existence."[13] The profound failure of the best system of regulation raises difficult issues concerning the relationship between conservation and development.

THE COLONIAL HISTORY OF THE NORTHWEST ATLANTIC FISHERY: EXPANSION INTO A SUBMERGED CONTINENT

> The cod appears to be one of the most prolific kind of fish.
> Of this there need be no other proof than the great number
> of ships which annually load with it . . . .
>
> Antonio de Ulloa (1758)[14]

> No other industry has engaged the activities of any
> people in North America over such a long period of time
> and in such a restricted area.
>
> Harold Innis[15]

Arising out of this plenitude of fish and human activity, Innis' monu-
mental work *The Cod Fishery: The History of an International Economy* (1940)
provides an interesting commentary on an industry that is now — on the
East Coast of Canada in the 1990s — in a state of almost complete ecologic
and economic collapse. In discussing the history of colonial exploitation in
the Northwest Atlantic over the last four centuries, Innis' work contributes a
historical context to the current environmental perspective related to the
political economy of depletion and dependence. Central to the correlations
between Innis' work and those based on depletion and dependence are the
relationships of international political economy which promote overexploitation in particular geographic realities and natural processes, while at the same
time fostering socio-political frameworks of dependence which leave local
communities in exceedingly vulnerable positions as they attempt to cope
with the aftermath of ecologic and social dislocation. It is this condition of
depletion and dependence which links the plight of Atlantic Canadian
coastal communities with that of local cultures in the Southern Hemisphere
as they struggle for survival in the context of overexploitation and international trade and debt arrangements.

For Innis, both political economy and natural community requires a
comprehensive analysis for there to be an understanding of the relationships
that exist in the fishery. Innis summarizes the Canadian interrelationship
between natural systems and economic history in this way:

> The economic history of the regions adjacent to the sub-
> merged areas extending to the northeast of America's north
> Atlantic seaboard is in striking contrast to that of the conti-
> nental regions. In the continent's northern area the St.
> Lawrence facilitated expansion westward and a concentration

on fur, lumber, and, and wheat; in the submerged areas innu-
merable small, drowned river valleys in the form of bays and
harbours facilitated expansion eastward and a concentration
on fish. Drainage basins bring about centralization, submerged drainage basins decentralization. . . . Unity of structure in the economic organization of the St. Lawrence was in
sharp contrast with the lack of unity in the fishing regions. . .
. In the interior, economic history was marked by changes to
new staple industries; on the Atlantic, changes were centred
in a single industry.[16]

In the view of later economists who had become enamoured by the adapt-
ability of capital and expanded forces of production, this recognition of
the importance of natural and geographic realities could appear to be an
almost Physiocratic linking of nature and economy. But from the perspective of the current collapse of marine communities in the Northwest
Atlantic, Innis' attention to natural processes is noteworthy, and is in sharp
contrast to the corresponding invisibility of ecological realities in the
works of later economists and managers — like the ones who were part of
the Kirby Task Force's study on the fishery *Navigating Troubled Water*
(1983) — whose economic priorities took little account of the limits of
the workings of natural communities, and were more concerned with creating large, vertically-integrated fish companies as a solution to the problems in the fishery. As Barbara Neis states with regard to Fordist
economies-of-scale generally:

The Fordist relationship between capitalism and nature was
based on seeking out, at a global level, large, dependable
supplies of relatively homogeneous raw materials such as oil
and wheat. In other words, Fordism relied heavily on direct
and indirect control of such natural resources by large
multinational corporations and relatively little knowledge
about nature and on the efforts to transform nature. . .[17]

Neis also points out that a great many theorists who attempt to analyze the transformation from Fordist to post-Fordist approaches to production "neglect the barriers to capital accumulation which nature imposes."[18]

The political economy of depletion and dependence returns analysis necessarily to the staple of the process without which nothing else can happen. It also counters the homelessness of capital in affirming the located sense of the relationship between human community and natural community — not only as operands of the production process — but in terms of their interrelated and situated contexts.

INNIS AND MARINE LIFE

Innis begins his study of the economic history of Northeastern North America with an in-depth discussion of the biological characteristics of the cod fish and the marine area of which it was a part. "An interpretation of [the cod's] significance in the economic history of the area depends on an understanding of its geographical background and habits."[19] A submerged analysis of underwater realities leads Innis to give details on water temperature, egg laying, available food for cod at different times of year, behaviour of small fry, and the effects of ocean current and wind direction on food availability and school migration. This recognition of the importance of natural realities locates economic activity within natural processes:

The great wealth and complex interdependence of animal life along the seaboard of the Maritimes have as yet baffled the scientist, and only small areas which have yielded to economic exploitation have come under the range of intensive investigation. The Banks are subjected primarily to ocean phenomena, and are not influenced by rivers from Newfoundland or by fresh water from the Gulf of St. Lawrence. The Gulf Stream and the Labrador Current, a variety of conditions of temperature and climate, and a food sup-

ply varying from plankton to the larger fish in the vicinity of
the Grand Banks are responsible for the abundance and
diversity of the animal life which supports the extensive but
fluctuating cod fishery.[20]

As well as pointing out the many natural characteristics which are of great
significance to economic activity in the fishery, Innis alludes to the fact that
"only small areas which have yielded to economic exploitation have come
under the range of intensive investigation." The importance of linking scientific study with economic activity in the fishery cannot be overestimated in
an examination of ecologic collapse. What Innis' statement points to is that
science operates in the service of economics, rather than in the service of biological conservation. This has certainly been the case in the history of the
fishery. The resource management myth that for each economic imperative
there is an equal and opposite regulatory response in the name of conservation, bares little relation to the events.

In his discussion of nature's economy, Innis discusses the specific
habits of the cod fish:

> The cod prefers a salinity of 34 per thousand and a tem-
> perature of 40 to 50 degrees, but its range is far beyond
> these limits. It frequents chiefly rocky, pebbly, sandy, or
> gravely grounds in general from 20 to 70 fathoms in
> depth, although it has been taken at 250 fathoms and
> thrives in temperatures as low as 34 degrees. The cod usu-
> ally spawns in water less than 30 fathoms deep and appar-
> ently in fairly restricted areas. A female 40 inches long
> will produce 3,000,000 eggs, and it has been estimated
> that a 52-inch fish weighing 51 pounds would produce
> nearly 9,000,000. The eggs float in the upper layers of
> water, where they are fertilized and hatched. . . .
> Experiments have shown that a temperature of 47 degrees
> will lead to hatching in 10 or 11 days, of 43 degrees in 14

or 15 days, of 38 or 39 degrees in 20 to 23 days, and of 32 degrees in 40 days or more.[21]

This is followed by a detailed anatomical description of the cod fish, and a discussion of the seasonal role of herring, capelin, squid, and crustaceans in the cod's diet. This analysis of the cod's habits and characteristics concludes by pointing out the ways in which these realities affect both the fishing activity of those engaged in the industry, as well as the curing and preserving methods which made the cod a tradable commodity — something which became important to the economic history of northeastern North America.

The attempt to include natural processes in economic analysis is currently considered to be of major importance in analyzing environmental problems and developing a perspective which would promote sustainable uses of natural communities. This is in contrast to the expansionary approaches associated with Fordist capitalism which took little or no account of the relationships in natural communities. Alternatively, Innis' work links human economy and nature's economy, despite the fact that there were few problems related to overexploitation when he was doing his research. The very real likelihood that the cod fishery will disappear in economic terms in Atlantic Canada, necessarily creates a new context in which to examine Innis' study into the central importance of the cod fishery to the economic history of the region. It also leads to the recognition of the centrality of the relationships in natural communities to the well-being of human communities.

INNIS AND INTERNATIONAL POLITICAL ECONOMY

Most of *The Cod Fishery: The History of an International Economy* is devoted to an analysis of the colonial relationships competing for dominance in Northeastern North America and the effect these struggles — and the accompanying trade and commerce arrangements which informed the colonial powers' exploitation of the cod — had on local political and economic realities, such as the development of responsible government. This desire of colonial powers such as the French, British, and Spaniards to use the cod in their three-way trade with colonies in the Caribbean, had the indirect result of limiting the development of the colonies directly adjacent to the fish. As Innis states:

> The activity of commercialism based in the fishing industry and the relative articles of shipbuilding and trade fostered by the navigation system had significant implications for constitutional development. . . . While such legislation coincided with the demands of West Country [of Great Britain] commercial interests, it clashed with the interests of the colonies under the Crown. . . . The problem of empires was one of constitutional as well as economic organization.[22]

This led to conflicts where

> The West Country opposed the formation of settlements in Newfoundland to the point of hastening the rise of the fishing industry in New England. Nova Scotia, in turn, resisted the control of New England and accentuated the isolation of Newfoundland. . . . Exports of sugar from the British West Indies led to the emergence of vested interests which fostered legislation opposed to the trade of New England and the colonies. . . . Direct trading between the West Indies and England flourished at the expense of the auxiliary trading between the colonies and the West Indies.[23]

Although "an expanding commercial system broke the bonds of a rigid political structure defended by vested interest,"[24] it was followed by a more directly capitalist arrangement after 1783 whereby "the new empire was more firmly based on direct exports to Great Britain in return for finished products, and the monopolies of the old empire became impossible because of the importance of trade with the United States."[25] These colonial arrangements weakened during the 19th Century, but this centre-periphery reality still exists today in the relationship between Atlantic Provinces and the Canadian Government, and undermines conservation measures because of the demand for economic growth in a perceived "have-not" region which still does not control economic and trade arrangements.

The context for the international exploitation of cod has continued to be important throughout the last fifty years and has played a significant role in the collapse of marine communities. These international forces have had both direct and indirect results. In direct terms, the unregulated international distant-water fleet — which included up to twenty industrialized nations — expanded dramatically after World War II and exploited the cod to the point of collapse in the early 1970s. Indirectly, the massive catching capacity of international factory freezer trawlers led to a lack of development in the Canadian fleet, which remained in large part artisanal. There was also a lack of development in accompanying regulatory frameworks within Canada while the industry became international, and — after the declaration of the 200-mile limit in 1977 and the Canadianization of the industry — this "underdevelopment" resulted in a confusion of expansion and restriction in the attempts to both regulate and increase the exploitation of a newly nationalized resource.

Innis' emphasis on the economic ramifications of the shift from a salt fishery to a fresh frozen industry, and Atlantic Canadian underdevelopment, is of special significance. By focusing on underdevelopment in the region and the increased pressure on marine communities brought on by changes in international economic relations, Innis' work in the 1930s describes the relationships which promotes the severe ecological problems which occurred for the first time in the late 1960s.

Innis gives this description of the shift from salt fish to fresh fish and the complex arrangements which were central to this shift:

> The spread of industrialism evident in urbanization, improved transport, and refrigeration had profound effects on an industry that had its life in a commodity which depended on salt as a preservative if its product was to be sold

in distant and tropical countries. . . . The overhead costs of large-scale equipment in the fresh-fish industry tended to force dried cod into the position of a by-product.[26]

Other changes in technology and communications accelerated this process of transformation which made fresh and frozen fish more readily available:

> . . . the broadening of the market in the United States, together with refrigeration and improved communications by telephone, telegraph, and radio, brought about improved facilities for handling fresh fish. . . . The introduction of the filleting process in 1921 and the marketing of packaged fillets reduced the weight of fish and expanded the market. . . . A rapid increase in trawlers accompanied an expanding market. . . . With the decline of the fishery on Georges Bank in 1931 [due to overfishing], there was a sharp increase on the other banks, which became more accessible due to more rapid steam and motor ships.[27]

The shift from the low investment levels needed for salt-preserving and long-line technology used on salt-bankers such as the *Bluenose*, to the high investment trawlers — or draggers as they are now called — and the consequent demand for a regular supply of fish which accompanied this increased investment is described by Innis:

> Trawlers, while not needing bait, require an abundance of coal and ice. They can support with greater dependability, and under a variety of weather conditions, a market demanding larger quantities of fish on certain days of the week and during certain seasons of the year. . . . the large-scale capital investment now essential to the fresh-fish industry — that is, an investment in cold-storage equipment, packing equipment, and by-products plant, extending in some cases to the

ownership of mills for the production of lumber — demands a continuous supply of raw material.[28]

This shift from salt to fresh fish intensified the pressure of capital because it converted the fish off Canada's East Coast from what had been a slave's food in tropical countries as part of the triangular colonial trade in the Atlantic, to a commodity in demand in the world's more prosperous countries:

> The shift from salt to fresh was also a shift from low standard of living countries to high standard of living countries. This increased the importance of capitalist modes of production and consumption.[29]

Innis' analysis of the colonial relationships which competed for the fish off the Northeast Coast of North America provide an important background for understanding the relationships which were involved in the events leading up to the ecological collapse of the cod. The post-colonial intensification of exploitation by the international fleet ended with the collapse of a range of Atlantic species in the early 1970s. The access the international fleet had to the fish also led to the limited development of the Canadian fleet and Canadian regulatory infrastructure. When the coastal zone was nationalized in 1977, this sole access and regulatory mandate proclaimed by the government of Canada led to a collision of the perspectives of expansion and regulation. This resulted in a situation where there was the expansion of Fordist industrial arrangements in the national context. There was an internalization and re-entrenchment of Canada's centre-periphery "colonial" relationships as represented by Ottawa on the one hand, and Atlantic coastal communities on the other. It is this history of depletion and dependence which now characterizes the current crisis in Canada's East Coast fishery, as well as many other local cultures in the Southern Hemisphere.

THE NORTHWEST ATLANTIC FISHERY (1945-1977)

Following World War II in the Northwest Atlantic off Canada's East Coast, the catches of the members of the International Commission for the Northwest Atlantic Fishery (ICNAF) — made up of the industrialized fishing nations of the world present in the area — rose from a level of 1,846,000 tonnes in 1946 to a high of 4,599,000 tonnes in 1975. (These statistics are from ICNAF Statistical Bulletins unless otherwise indicated.) In the case of the groundfish off Canada's East Coast (which I am dealing with specifically), catches peaked in 1968 at 2,780,000 tonnes. The index of fishing effort, which is a combination of boat tonnage and days fished, expanded by a factor of twenty between 1954 and 1974. The fishery's transition from dory-vessels and salt fish, through side-trawlers, to the now predominant stern-trawler, was characterized by an increase in catching capability that converted a labour intensive industry into a capital intensive industry.[30]

As discussed above, Innis chronicled the economics and the technology of the salt cod fishery in the Northwest Atlantic. It was based on setting dories off from a sailing vessel to handline with a single hook for demersal, or groundfish (as they are called) that feed at or near the ocean bottom, and which include cod, haddock, cusk, hake, pollock, flounder, and halibut. The day's catch would be split and salted in the fish hold of the vessel. It could take months to load the vessel to its 300,000 pound capacity. Much of this fish returned on the Portuguese and Spanish fleet to Europe or was traded in the Caribbean for rum and molasses as part of the triangular colonial trade which used salt cod as slave food for plantation labour.

With the development of the beam trawl, these later diesel-powered vessels were converted into side-trawlers. Rather than catching the groundfish with hooks as had been done in the past, a funnel shaped net was towed along the ocean bottom. The forward end of the net into which the fish entered was held open by a long steel beam and then later by trawl doors angled outwards. This kind of fishing was made possible by the increased power and speed of diesel engines, as well as by the development of hydraulic winches and stronger synthetic fibres used in the net trawls.

Existing vessels, with their wheelhouse aft, were retrofitted with this technology and the net was hauled in over the side of the vessel. The number of side-trawlers fishing in the Northwest Atlantic increased from 560 in 1953 to 1316 in 1965. Their numbers decreased dramatically after that as they were replaced by the more efficient and powerful stern-trawler, which was specifically designed to use the new trawl technology. By 1974, stern trawlers represented three quarters of the gross registered tonnage of vessels fishing off Canada's East Coast. To quote D. A. Pepper:

> . . . it is accepted that the stern trawler was vastly more effi-
> cient over the older side-trawler and it was this newer vessel
> that permitted an increased range of operations.[31]

These new stern-trawlers were larger than the side-trawlers with more powerful engines. The wheelhouse was moved to the bow to create more workable deck space as well as to accommodate the stern ramp and gallows which made it possible to haul a much larger net up the stern of the vessel. It also allowed the fleet to fish on more inclement days because the boat could be kept head to the waves and wind while the net was being hauled aboard, as opposed to the side trawler which had to lay 'side-to' when hauling the net. Given the long distances travelled by the distant-water fleet and the variability of the weather on the Northwest Atlantic, features such as these greatly increased the effectiveness of this catching practice. Along with the increased effectiveness of each vessel, the total number grew from 620 in 1954 to 1537 in 1974, and tonnage increased during this period by 500%. These transformations led to an exponential growth in the index of fishing effort (a combination of days fished and gross registered tonnage) from 13,280 in 1954 to 241,453 in 1968. Fleets such as the Russians integrated their operations so that trawlers supplied fish for factory ships which were in turn serviced by supply ships that ferried fuel and new crews back and forth to the fishing grounds. To quote Pepper again:

The tremendous build up of the various high seas fleets of the ICNAF members produced an increase in landings and fishing effort. However, the lack of control resulted in the classical pattern of over-fishing.[32]

Pitcher and Hart echo a similar view of the recent events in the Northwest Atlantic:

> In the 1960s this area bore the brunt of the development of European distant-water fleets. The fishing pressure that developed greatly reduced the stocks of cod, haddock, and herring so that at present they are considered to be depleted.[33]

Catch levels dropped dramatically in the Northwest Atlantic during the early 1970s. The distant-water fleet fishing in the area responded to these drops in the traditional catches of cod and haddock by shifting their attention to less exploited species such as redfish while expanding the area in search of less intensely fished grounds such as the Labrador Sea.

This overexploitation of marine biotic communities took place in an unregulated, international context. ICNAF had only minimal regulations concerning resource exploitation and even these it could not enforce. Up until 1975, restrictions concerned mainly gear type, and only after the collapse of marine biotic communities in the early 1970s did ICNAF attempt to impose country-by-country catch quotas. Because it was unable to enforce these quotas in an international context, Canada, along with other coastal states, felt called upon to unilaterally declare a 200-mile coastal economic zone in 1977 in the name of conservation.

CANADA'S EAST COAST FISHING INDUSTRY

During this period of intensification in the industrialization of the fishery in the Northwest Atlantic, the Canadian fleet's share of the catch gradually dropped to a low of 20% in the mid-1960s. In response to this expansion of international catching capacity, the Canadian government, as well as Atlantic provincial governments, encouraged the development of a domestic offshore fleet to compete with other ICNAF members through a series of subsidies, low interest loans, and the granting of new licenses. As the fish were depleted by this increase in catching capacity, the traditional inshore fleet of small boats which counted on the migration of schools of fish close to shore, and which had in the past caught a majority of the fish, was now being threatened. In response to this polarization of the fishing fleet, the Canadian government continued to call for increases in "rationalization" and "efficiency" of the fishing industry. This accelerated the expansion of the domestic offshore fleet and increasingly portrayed the inshore fleet as an intransigent impediment to this process of modernization. To quote Pross and McCorquodale:

> Canada's first policy reaction to the post-war invasion of for-
> eign fleets was to develop her own competitive capacity. From
> the beginning of the period Ottawa was preoccupied with
> what it felt was the backwardness of the fishing industry.[34]

Prior to 1977, Canada had so little control over the competition for the fish off its coast, that much of its involvement in the fishery took the form of income stabilization. Especially with regard to the inshore fishery, much of the fisheries policy took the form of social policy. In 1965 for example, this approach led to such drastic social measures as the agreement between the federal government and the government of Newfoundland to move people from outports to more centralized locations, where it was hoped there would be more employment and an upgrading of services. This move also contributed to the labour force for the development of the offshore fishery which was to be concentrated in these new relocation centres.

In an attempt to overcome what it saw as a stagnant economy based on a traditional inshore fishery, Canada embarked upon the wholesale development of a sophisticated offshore industrialized fleet. This process is summed up by Pross and McCorquodale:

> In order to establish a modern, efficient fishing industry with financial returns comparable to other industries, it was recognized that the individual fisherman would have to modify substantially his catching techniques. Federal officials felt that the greatest opportunities for individual benefits lay in the offshore industry. The large volume and regularity of offshore landings could support a fresh/frozen industry onshore. The many limitations, man-made and natural, on the productivity of the daily, inshore fishing operation made that sector unreliable for a dependable supply of raw material for the fresh/frozen processing facilities. Therefore, federal policy was designed to encourage fishermen to shift into offshore fishing operations which utilize large vessels and mechanical gear. The objective was pursued largely through subsidies for vessel construction which fostered the expansion and modernization of the Atlantic offshore fleet.[35]

Between 1947 and 1960, 125 longliners and 34 draggers were constructed in Nova Scotia with government assistance. In 1953, the federal government ensured that development money went to the offshore sector by passing a regulation that made subsidies contingent upon the affiliation of large trawlers with processing companies.[36] As a result of these programmes, the number of Canadian vessels over 50 tons increased by 165% from 211 to 558 and the total tonnage of the fleet increased by 320% during the 1960s.

Nowhere is this relationship between development and government policy more intertwined on Canada's East Coast than in the growth of the large vertically-integrated offshore companies such as National Sea Products. During the period of modernization and industrialization, National Sea became "the largest organization on the Atlantic Coast of North America engaged in the production and processing of fish."[37] In an attempt to regain some of the catch lost to the expanding foreign fleet, both federal and provincial governments subsidized the construction of vessels and processing facilities for offshore companies such as National Sea. Set up initially after World War II, National Sea's trawler fleet expanded to 44 vessels by 1964, thereby becoming the largest on the Atlantic Coast of North America. In that same year, the company constructed the largest processing facilities in North America at Lunenburg, Nova Scotia. government aid for the plant totalled 3.5 million dollars. In spite of the fact that catches dropped sharply after the peak in 1968, government support for the expansion of National Sea continued throughout the 1970s:

> Despite consistent losses on the operations of its offshore trawler fleet, NSP was able to expand, innovate and intensify its operations at a point of resource depletion. This was possible only through three forms of state subsidies. Direct capital grants, for instance, totally subsidized the construction costs of new trawlers in 1968. Such grants were used by the company to increase already sizable capital cost allowances under income tax regulations, while deferred taxes provided a consistent source of working capital throughout the period. The state's role in such development programs seemed to contradict completely its resource management function.[38]

During its peak period of expansion in the early 1970s, government support for National Sea equalled one-quarter of the company's total annual budget. Barrett sums up the history of the relationship between the government and National Sea Products in this way:

> The history of National Sea Products is one of growth and expansion under the protective wing of a developmentist state, especially up to the 1970s. In payment for this public tutelage, the company took advantage of every opportunity to exploit underutilized species or new species of fish, and to

expand efforts into more traditional fisheries. Centralism, concentration and technological modernization became its hallmarks. In spite of this seeming orderly expansion, however, anarchy and frenzied overexploitation prevailed. When fish stocks were threatened, the company could only respond by increasing efforts in other areas or by diverting capital out of the fishery or out of the country altogether. To such an organization conservation and rational management were an anathema.[39]

As catch levels dropped in traditional species such as cod and haddock, government money underwrote the expansion of exploitation to "underutilized species" such as scallops, redfish, and snow crab, as well as to new fishing areas such as the Labrador Sea. Although government support of industrialization continued throughout the 1970s, Canada's share of the catch in the Northwest Atlantic continued at low levels, and in spite of the 320% increase in the gross tonnage of the offshore fleet in the 1960s, catch levels only increased by 18% during this period. This eventually led to Canada's declaration of the 200-mile limit in 1977, in hopes that exclusive domestic access to the fish would increase Canadian landings and conservation measures would be put in place to protect biotic communities. The hoped-for prosperity that was to follow the declaration of the 200-mile limit did not come to pass. The catching capacity of the international fleet which had caused overexploitation was merely replaced by the newly constructed Canadian fleet, albeit operating in a more closely-regulated fishery. This industrial expansion led initially to a crisis based on overproduction, large inventories, and excessive debt during the early 1980s. This was investigated by the Kirby Task Force (1983) which recommended the restructuring of the offshore sector. Later in the decade, this same overcapacity in the catching sector caused a serious depletion of the fish, as set out by the Scotia-Fundy Groundfish Task Force (1989), which called for a scaling down of the fleet.

This kind of development has left the East Coast fishery with an overcapacitized and therefore inefficient offshore fleet, and a less debt-ridden inshore fleet that nonetheless has few fish to catch and requires ongoing income supplements. In either case, government money was very much at the centre of both company subsidies and income stabilization for individual fishermen. These financial commitments to the participants in the industry greatly exacerbated the federal government's new founded role as resource manager in charge of conservation after Canada nationalized coastal waters in 1977.

In attempting to come to terms with what caused the ecologic collapse of the cod under a national management regime, and which led to the fishing moratorium being declared in July of 1992, the *Task Force on Incomes and Adjustment in the Atlantic Fishery*, chaired by Richard Cashin, stated in their report *Charting a New Course: Towards the Fishery of the Future* (1993):

> The fundamental problems in the fishery may be summed up
> as three elements — overdependence on the fishery, pressure
> on the resource, and industry overcapacity — all interacting
> in a vicious cycle.[40]

It is clear that the forces which are causing the collapse of a range of fisheries across the globe are all but identical with the forces which caused the collapse of Canada's East Coast fishery. These forces have created a crisis which reflects the general problem of ecologic brinkmanship:

> Given the greatly reduced groundfish quotas, fisheries clo-
> sures and poor catch performance to date, the projected 1993
> groundfish catch will be no more than 250,000 tonnes.
> Compared with 1982, this means that the groundfish base of
> the Atlantic Canadian fishery will have shrunk by more than
> 500,000 tonnes.[41]

The commitment of national governments to large-scale, highly-industrialized approaches to fishing, has contributed to both the undermining of traditional fishing sectors and the natural communities on which they have depended. Of course, traditional fishing in the Atlantic Canadian context can only refer to aboriginal fishing, which was undermined first by colonial relations and then by international commodity relations.

Inspite of the profound failures that have plagued the industry, policy-makers and resource planners resist the recognition that modern economic imperatives are at the heart of this economic and ecologic dislocation. Although Cashin states that, with regard to the ecologic collapse in Canada's East Coast fishery, "We are dealing here with a famine of biblical scale — a great destruction," and although all the evidence points in that direction, the report is at pains not to point the finger for this destruction directly at modern economic and technological realities. The Cashin Report states: "What has caused or contributed to this unprecedented and widespread resource collapse? There is no definitive evidence, but there are a number of factors which, in varying degrees and combinations, have a role in this decline."[42]Other reports in the aftermath of the collapse of the fishery are not so circumspect. In a 1994 report for the Federal Department of Fisheries and Oceans, Hutching and Myers state:

> The temporal changes in demography, population sustain-ability, harvest rates, and inshore/offshore catch rates docu-mented here provide strong evidence that overexploitation was the primary cause of the collapse of the northern cod in the early 1990s.[43]

What the spectre of ecologic collapse confirms is that modern market economy, coupled with technological capability, has decimated a wide range of marine communities in the fishery. Farley Mowat describes this failure in viability in terms of a "financial vortex [of] . . . massive high-tech fishing for profit."[44] It is this vortex of modern economy which regulation, as defined in the context of resource management and sustainability, attempts to bring under control.

NOTES

1   Bruce Mitchell. 1989. *Geography and Resource Analysis*. London: Longman & Wiley, p. 5.

2   Gustavo Esteva. 1992. "Development." in *The Development Dictionary* [Wolfgang Sachs ed.] London: Zed Books, p. 18.

3   Pitcher, T. J. & Paul Hart. 1982. *Fisheries Ecology*. London: Croom Helm, p. 49-51.

4   Cushing, D. H. 1987. *The Provident Sea*. New York: Cambridge University Press.

5   Peter Weber. 1995. "Protecting Oceanic Fisheries and Jobs." in *State of the World 1995*. [Lester Brown ed.] New York: Norton, p. 21.

6   Weber. (1995:24).

7   Weber. (1995:24).

8   Weber. (1995:24-25).

9   Weber. (1995:27).

10   Weber. (1995:25).

11   Peter Weber. 1993. *Abandoned Seas: Reversing the Decline of the Oceans*. Washington: Worldwatch Paper 116, p. 32-33.

12   James R. Coull. 1993. *World Fisheries Resources*. London: Routledge, p. 176.

13   Coull (1993:176).

14   Quoted in: Harold Innis. 1954. *The Cod Fishery: A History of an International Economy*. Toronto: University of Toronto Press, p. 2.

15   Innis. (1954:2).

16   Innis. (1954:484).

17   Barbara Neis. 1993. "Flexible Specialization: What's That Got To Do with the Price of Fish?" in *Production, Space, Identity: Political Economy Faces the 21st Century*, [Jane Jenson, Rianne Mahon, and Manfred Bienefeld eds.]. Toronto: Canadian Scholars Press Inc., p. 90.

18   Neis (1993:88).

19   Innis (1954:1).

20   Innis (1954:2).

21   Innis (1954:3).

22   Innis (1954:506).

23   Innis (1954:500&501).

24   Innis (1954:502).

25   Innis (1954:502).

26   Innis (1954:418-419).

27   Innis (1954:423).

28   Innis (1954:435).

29   Innis (1954:443).

30   Pepper, D. A. 1978. *Men, Boats, and Fish in the Northwest Atlantic.* Cardiff: Department of Maritime Studies, University of Wales.

31   Pepper (1978:259).

32   Pepper (1978:271).

33   Pitcher & Hart (1982:54).

34   Pross, A. P. & S. McCorquodale. 1987. *Economic Resurgence and the C onstitutional Agenda: The Case of the East Coast Fisheries.* Kingston: Queen's University, p. 12-13.

35   Pross & McCorquodale (1987:13-14).

36   Barrett, Gene. 1984.

37   Barrett (1984:86).

38   Barrett (1984:88).

39   Barrett (1984:96).

40   Richard Cashin. 1993. *Charting a New Course: Toward the Fishery of the Future.* Ottawa: Minister of Supply and Services, p. 14.

41   Cashin. (1993:20).

42   Cashin. (1993:21).

43   Jeffrey A. Hutchings & Ransom A. Myers. 1994. "What Can Be Learned from the Collapse of a Renewable Resource?: Atlantic Cod, *Gadus morhua,* of Newfoundland and Labrador." St. John's: Science Branch, Department of Fisheries and Oceans, p. 39.

44   Farley Mowat. 1990. Quoted in "Hibernia Blues" by G. Wheeler. *Now Magazine.* Vol. 10, No. 4, Sept. 27 - Oct. 3, p. 10.

III

Resource Management:
From Common Property to Public Property
to Private Property

. . . Resource management represents the actual decisions
concerning policy or practice regarding how resources are
allocated and under what conditions or arrangements
resources may be developed.

Bruce Mitchell[1]

. . . the Progressives assumed that lawmaking was somehow
divorced from competition in the marketplace and not . . .
in many respects a struggle for resources carried on by
other means.

Arthur McEvoy[2]

In a discussion paper leading up to the *Law of the Sea
Conference* in 1974, the Canadian Department of the
Environment stated that it was in favour of the 200-mile
Exclusive Economic Zone because, "the Canadian Government
considers customary international law inadequate to protect
Canada's interest in the protection of the marine environment
and its renewable resources."[3] In response to the inadequacy of
international law, and on the eve of the declaration of that 200-
mile zone, the Canadian Government layed out its first comprehensive fisheries initiative in *Policy for Canada's Commercial
Fisheries*. With regard to regulating the exploitation of fish, its
stated goals were:

- Obtain national control of the exploitation of fishery resources throughout a zone extending at least 200 nautical miles from the Canadian coast.
- Secure international recognition of the state of origin's primary interest and responsibility for anadromous fish species.
- Provide for redevelopment and enhancement of fish stocks whose natural habitat or environment is amenable to effective modification.
- Institute a co-ordinated research and administrative capability to control fishery resource use on an ecological basis and in accordance with the best interests (economic and social) of Canadian society.
- Provide the research and the institutional innovation necessary to foster the development of viable aquacultural enterprise.
- Allocate access to fishery resources in the short-run on the basis of a satisfactory trade-off between efficiency and dependency of the fleets involved.
- Develop a fully effective capability for the monitoring of information on resource and oceanic conditions, for the surveillance of fleet activity and for the enforcement of management regulations.[4]

Compare this with the stated goals of the *World Conservation Strategy*, published four years later:

- Determine the productive capacities of exploited species and ecosystems and ensure that utilization does not exceed those capabilities.
- Adopt conservative management objectives for the utilization of species and ecosystems.
- Ensure that access to a resource does not exceed the resource's capacity to sustain exploitation.

- Reduce excessive yields to sustainable levels.
- Reduce incidental take as much as possible.
- Maintain the habitats of resource species.[5]

It is clear that the stated objectives of both policy papers are very similar in their approach to integrating the forces of development with conservation initiatives. It is therefore the purpose of this chapter to chart the course of the failure of conservation measures in practice in a way that contributes to the theoretical discussions of sustainability. In doing so, I will focus on the links between the evolving conservation strategies of the Northwest Atlantic fishery and the evolving strategies for global conservation. Central to this discussion will be an analysis of the relationship between regulation and exploitation, beginning with the international open-access scenario, then through the period of comprehensive national management, and concluding with initiatives which privatize the fish in the ocean. There is, therefore, a close relationship between property rights and management approaches. This focus on property rights provides an interesting avenue into the rationale for management. It also highlights the assumptions which underwrite management of the natural world in a modern economy, and sheds light on the failures of conservation.

The previous chapter focused on exploitation in the international open-access situation which eventually led to ecologic collapse in the early 1970s. With the declaration in 1977 of the 200-mile limit in the Northwest Atlantic, what had been open access exploitation by the international community was now under Canadian jurisdiction. Although foreign countries still had access to the fish inside the 200-mile limit under prior agreements, the Canadian government had the mandate to set quotas as well as gear and vessel restrictions.

As it developed the organizational infrastructure to manage this publicly-owned benefit on behalf of the people of Canada, the federal government was conscious of its responsibilities, both to conserve the fish, and to foster orderly economic growth in the fishing industry. A complex web of regulations to control access to the fish gradually evolved, as well as a series of incentives for the development of the industry. It was hoped that with increased involvement in the fishery, the federal government could increase the net benefit derived from the fish, both for those who participate in the industry and for the Canadian taxpayers as a whole who fund its regulation.

In order to provide this orderliness of access to the fish, as well as to insure conservation, the Canadian government embarked on a program of limited-entry licensing, gear and vessel restrictions, and the extension of private property rights. The goal was to control exploitation so as to maximize the benefit to those involved in the fishery, as well as promote the conservation of biotic communities. Despite these measures, fleet and processing overcapacity occurred, leading to the overexploitation of the fish.

Because fishery managers and policy-makers held that modern market economy was "normal business" practice and because the one component in the fishery that did not fit the "normal business" model was the lack of private property rights, the chronic problems in the fishery were ultimately blamed on the lack of property rights. In other words, the problems in the fishery were attributed to the fact that it was not capitalist *enough*. Therefore, most of the policy initiatives in the fishery were focused on implementing a comprehensive regulation of public property, followed by the granting of private property rights. This same market-based approach has come to dominate the sustainability debate as well, and is reflected in the proposal to trade pollution permits to the industry or in effluent taxes. This privatized market-based approach is seen as a more efficient way of solving the fishery's problems than the regulatory approach which is expensive for governments and causes conflict between the private and public sector.

COMMON PROPERTY, OR GLOBAL OPEN ACCESS, IN THE NORTHWEST ATLANTIC FISHERY

Canada's East Coast fishery have long suffered from what is understood as the problems of common property resource exploitation. A theory that incorporated the common property aspect of the fishery's resource management, was first articulated by Scott Gordon in the article published in 1954 and entitled, "The Economic Theory of a Common Property Resource: The Fishery," Gordon states,

> Fishery resources are unusual in the fact of their common property nature. . . where natural resources are owned in common and exploited under conditions of individualistic competition.[6]

Summarizing the research done on the fishery previous to the publication of his work, Gordon concludes,

> Virtually no specific research into the economics of fishery resource utilization has been undertaken. The present state of knowledge is that a great deal is known about the biology of the various commercial species but little about the economic characteristics of the fishing industry.[7]

This has resulted in a situation where the biological concept of Maximum Sustainable Yield is the primary focus, with little regard for other factors in the industry:

> Focusing attention on the maximization of the catch neglects entirely the inputs of other factors of production which are used up in fishing and must be accounted for as costs. . . . In fact, the very conception of a net economic yield has scarcely made any appearance at all. On the whole, biologists tend to treat the fisherman as an exogenous element in their analytical model, and the behaviour of fishermen is not made into an integrated element of a general and systematic "bionomic theory."[8]

It is this "general and systematic 'bionomic theory'" which universalizes modern economy as the basis upon which fishery issues are analyzed. And it is the "behaviour of fishermen" which is entirely linked to modern economic rationality. So in applying economic theory to the fishery's common property focus, Gordon initiates a process whereby resource economists and policy-makers begin to assume the very things that need to be explained when it comes to the destruction of the natural world.

In developing his theory of common property resource exploitation, it was Gordon's goal "to demonstrate that the 'overfishing problem' has its roots in the economic organization of the industry."[9] The ramifications of this in terms of fisheries management was an increased responsibility on the part of government to regulate the fishing industry, not just in biological terms, which at this time was not considered a problem, but also in social and economic terms. This would lead to a management scheme where

> the optimum degree of utilization of any particular fishing ground as that which maximizes the net economic yield, the difference between total cost, on the one hand, and the total receipts (or total value production), on the other.[10]

Gordon concludes that the open access nature of the common property fishery leads to overcapacity which leads to depletion of both income and fish, with little economic benefit being accrued:

> There appears, then, to be some truth in the conservative dictum that everybody's property is nobody's property. Wealth that is free for all is valued by none. . . . The blade of grass that the manorial cow herd leaves behind is valueless to him, for tomorrow it may be eaten by another's animal. . . . Common property natural resources are free goods for the individual and scarce goods for society. Under unregulated private exploitation they can yield no rent; that can be accomplished only by methods which make them private

> property or public (government) property, in either case sub-
> ject to a unified directing power.[11]

It is the reliance first on the "unified directing power" of government authority, and then on private property rights, which defines the history of Canada's
management of the fishery. It was government power or private property
rights which were supposed to rid the fishery of its instability and inefficiency. It is also government power or private property rights which form the
basis of a great deal of conservation measures as they are defined within the
sustainability debate. It is the failure of government power and private property rights to promote the conservation of the fishery which defines the history of Canada's involvement in the Northwest Atlantic fishery. It is this failure which presents a profound challenge to current discussions about the
relationship between conservation and development.

THE ATLANTIC CANADIAN FISHING INDUSTRY

While the British North America Act explicitly assigns jurisdiction
over fisheries to the federal government, judicial interpretation and administrative practice have somewhat eroded Ottawa's primacy in this area.[12]
Throughout the history of the fishery, all of the Atlantic Provinces have
assumed an active role in the fisheries administration in co-operation with
the federal government.

Up until the late-1960s, the federal government was content to pur-
sue a largely ad hoc approach to fisheries policy, responding to proposals
made by the industry, scientists and provincial governments. Many of the
measures to improve the fishery came in the form of subsidies and bailouts.
Although conservation was a federal responsibility which would have the
most widespread ramifications for the industry, previous to the 1977 declaration of the 200-mile limit, this function was administered by the
International Commission for the Northwest Atlantic Fishery (ICNAF), an
umbrella agency made up of countries that fished in the Northwest Atlantic.

As a result, federal responsibilities gravitated toward social policy, such as unemployment insurance for fishermen, and industrial development based on subsidies, rather than the conservation of fish.

Standard common property models such as Gordon's, show that under open-access conditions, where everyone who wants to fish can fish, there is a tendency to dissipate the economic value of the fish through the creation of excess capacity. Increasingly more labour and equipment are committed to catching the same amount of fish. These factors, it is argued, have contributed to instability in the industry and a waste of caught fish, especially when these economic realities are seen as operating in conjunction with biological factors such as fluctuations caused by changes in water temperature, food supply and migration patterns. The combining of these factors can result in a steep fall or rise in earnings and profits. Industry reaction to this form of uncertainty frequently has been to install sufficient catching and processing capacity to handle the peaks in supply, thereby inflating industrial overheads and reinforcing the inherent tendency toward over-expansion in the commercial fisheries. This leads to the kind of government policy that maintains fishermen and fish plants during downward cycles in an overcapitalized industry. This perpetuates the overexploitation of fish, as well as low incomes and government dependence during the periods of fluctuation in fish availability.

The government's response to the fisheries crisis in the early 1970s was an illustration of this kind of approach. With the rise in the price of fuel due to the OPEC oil embargo and the recession that followed in the United States (which softened the market for fish products), the federal government stepped in with a $14.5 million scheme for inventory financing and product promotion. It also purchased $1.5 million worth of canned products to give to international food aid.[13] Because of the increased costs of exploitation due to the increased cost of fuel (from 6 cents to 12 cents/pound), the government also instituted a series of payments per pound of first quality fish caught so as to shore up the fleet sector. This series of payments during the early to mid-1970s amounted to over $2,200 per fisherman.

As well as the many financial problems faced by the fishing industry, there was also an acute shortage of fish caused mainly by foreign overfishing in the early-1970s. During this period, the Canadian fleet was only catching about 20% of the groundfish off the East Coast. Thus, as well as having economic and biological dynamics which made the industry unstable, it was also vulnerable to fishing pressure from foreign fleets.

Up until the early-1970s, the only fishing regulations concerned gear type. There were no controls on the amount of fish caught. Because of the ecological crisis during the early-1970s, ICNAF attempted to establish country by country quotas in the offshore fisheries. The quotas, including that of Canada's offshore fleet, were reduced from year to year as a result of accumulating evidence of reduced abundance of fish. However, there were repeated problems concerning the enforceability of these regulations, both because of the international context in which they existed, and the volunteer nature of ICNAF. It is this same voluntary, international context for regulation which limits the effectiveness of so many of the United Nations conservation initiatives as they attempt to deal with interconnected environmental problems such as climate change, which are created by the economic priorities of individual member states.

Since they are initiated in the aftermath of a perceived problem, international regulation is confronted with the twin difficulties of ascertaining, in quantitative terms, the exact nature of the problem in terms of the levels of toxins or exploitation in an already volatile situation (both economically and ecologically), as well as attempting to ratchet down the demands of those already involved. In other words, it has difficulty in projecting a collective sense of a shared problem which can supersede individual interests. Hence the recurrence of the emphasis on the "common" significance of these problems which need to be addressed by wary competing members. The deeply entrenched paranoia of the Spanish fleet — expressed in the repeated statements which say that Canada was not motivated by a common concern for conservation, but was making a grab for more fish — reflects the general difficulty of promoting a common approach in a competitive market. It becomes a kind of environmental "Cold War" which poses a serious threat to the Earth's future. No one will give up standard practice, and therrefore everyone is threatened.

POLICY FOR CANADA'S COMMERCIAL FISHERIES (1976)

In response to these regulatory problems, the federal government's Department of the Environment issued Canada's first comprehensive fisheries policy, *Policy for Canada's Commercial Fisheries* in 1976. To quote from the policy paper,

> The crisis of the present, the chronic problems of the past, the expectations of national jurisdiction over offshore fishery resources in the future, the evidence of a potential for development of a viable fishing industry and the deepening financial involvement of government, all combined to bring matters to a head. The Minister of State for Fisheries, late in 1974, inaugurated a comprehensive study into the whole field of fishery management and development in Canada, with particular reference to revitalization of the groundfisheries of the Atlantic region.[14]

It is interesting to note that although there was an attempt to create a comprehensive policy, it was issued under the Department of the Environment, whereas the Fisheries portfolio was still a junior post of the Ministry of State. The comprehensive goal of the policy was not as yet reflected in terms of unified institutional responsibilities.

Recognizing the gravity of the situation, the government goes on to state that:

> The analysis of the fisheries carried out in 1974-75 by the Fisheries and Marine Service showed that, in some respects at least, fundamental restructuring of the fish-

ing industry is inevitable. It will come about either in
an orderly fashion under government auspices or
through the operation of inexorable economic and
social forces.[15]

Central to the new policy was the idea that the resource would no longer be
exploited at Maximum Sustainable Yield, but instead, quotas would be set at
a level which provided "the best use" of the fish for the "social benefits (personal income, occupational opportunity, consumer satisfaction and so on)
derived from the fisheries and the industries linked to them."[16] Combined
with this new, lower level of exploitation, the government was going to abandon past policy approaches which it defined as "simplistic in the approach to
resource management and relatively non-interventionist and unco-ordinated
in regard to industrial and trade development,"[17] and take on a role of
directly shaping the fishery in conjunction with industry interests.

Mirroring Gordon's analysis of the economic dynamics which govern
common property exploitation, the policy statement addresses the open
access approach that had prevailed up until this point:

In an open access, free-for-all fishery, competing fishermen
try to catch all the fish available to them, regardless of the
consequences. Unless they are checked, the usual consequence is a collapse of the fishery: that is, resource extinction
in the commercial sense, repeating in a fishery context "the
tragedy of the commons".[18]

This competition between fishermen in conjunction with open access,
generates overcapacity in the fleet, which in turn results in chronic economic problems for the industry, as well as the depletion of fish communities. Once again echoing Gordon with regard to regulation by a "unified directing power," but not, interestingly, in terms of private property
rights (that would come later), the policy paper states, "In both cases
what seems needed is a system of allocation by a body governing access

to the resource."[19] With regard to private property rights, the policy initiative argues:

> Since the establishment of private-property rights in fishery resources is impracticable in the great majority of cases, the state's responsibility for resource conservation and allocation cannot be delegated.[20]

This lack of regulation in the context of common property leads to "an unchecked scramble for advantage" which leads participants to the "verge of collapse."[21] In response to this "unchecked" behaviour, "Stability, with equitable access to a reliable resource, normally is maintained or restored only through the imposition of controls by government."[22]

With the declaration of the 200-mile limit on the horizon, many of the new policy directives were aimed at the Canadianization of the industry, including the development of a domestic distant-water fleet to replace the foreigners, the financing of conservation measures (called resource maintenance) by the industry itself, market development, quality control, and the setting up of the land-based industry to support this expansion. Standing in the way of this rationalization and the diversification of the industry was the lack of "acceptable alternative opportunities" for those people who, due to industry overcapacity, were trapped at a subsistence level of income. The 1976 policy paper outlines its goal this way:

> One requirement for a viable and prosperous commercial fishery is that fewer people be employed in relation to output in primary production. This does not mean drastic dislocation of the people now dependent on the fishing industry. It does mean that where it is feasible to expand, this expansion should be accomplished without increasing employment in the fishery itself. With a viable fishing industry firmly based and growing, new job opportunities

in a variety of associated industries and services would develop throughout the region.[23]

With the increased stability in the fishing industry brought about by these policy initiatives, combined with the expansion of access to the fish resulting from the declaration of the 200-mile limit, the government hoped that the spill-over effect of a larger and more prosperous fishery would create job opportunities in other sectors of the economy and solve some of the overcapacity problems in the fishing industry.

Central to the development of this more prosperous and stable industry was the government's goal of ending the open-access nature of the fishery. Along with the lack of employment alternatives in the region, what had always stood in the way of limiting open access was that, in the context of nationalization, fish has always been a common property owned by Canadians, and therefore all Canadians should have access to it. But with the increased role played by the federal government in the fishery, all Canadians were contributing a lot of tax dollars to maintain an industry at a low level of profitability. So, it was argued, because of its larger responsibility to taxpayers, the federal government was now instituting a policy of limited-entry licensing which it hoped would increase its profitability and create a larger net benefit to Canadians generally by limiting access to the fish both on the short and long-term basis:

> If Canadian society as a whole is to get the best combination
> of benefits from development of the fisheries, open access to
> resource use must be curtailed. . . . Effective entry control
> helps to stabilize the resource base and to smooth out the
> cyclical peaks and valleys on cost/price charts.[24]

In order to achieve this, the federal government will "institute a co-ordinated research and administrative capability to control fishery resource use on an ecological basis and in accordance with the best interests (economic and social) of Canadian society."[25] The mandate of this expanded infrastructure would include limiting entry to the fishery, enhancing the processing sector, consolidating the marketing of fish products, and developing community enhancement programs.

It was this multi-faceted approach to comprehensive regulation by the "unified directing power" of a national government which defined the mandate of Canada's declaration of the 200-mile limit and which is embodied in *Policy For Canada's Commercial Fisheries*. Conservation, as defined by a public regulatory body, would provide an equal and opposite response to fluctuating, unstable, and unsustainable economic competition which had preceded it. By using the latest and most sophisticated scientific means for assessing the size of marine fish communities, combined with a regulatory framework backed by the power of the national state, the Canadian government set out to fulfill the twin mandates of ecologic sustainability and economic stability in the Northwest Atlantic.

EXTENDED JURISDICTION (1977-81)

The development of the international distant-water fleets after World War II had jeopardized the domestic fishing activity of many countries around the globe. National rivalries, opposing interests, and the lack of enforcement rendered the conservation measures of international agencies ineffective. Originally put forth by Chile and Peru in 1947, the idea of a 200-mile coastal economic zone gradually gained acceptance by coastal states. Canada unilaterally declared its own 200-mile limit on January 1, 1977, along with other coastal states. At that point, the majority of the fishing grounds in the Northwest Atlantic became a part of Canada.

When Canada unilaterally declared its 200-mile economic zone in 1977, it nonetheless inherited a fish management plan from ICNAF. This commission was set up in 1949 and by the mid-1950s, was collecting catch and effort statistics as well as taking the first steps to regulate gear use (mesh size of dragger nets, for example). The biological criterion they used to set catch limits was called Maximum Sustainable Yield (MSY), which attempted to ensure that no commercial fish protein was wasted. Canada played a leading role in the organization, both in terms of biological research and technological developments relating to gear.

With catch levels reaching their height in the late-1960s, there was a general consensus that ICNAF was working well, although there was a great deal of difficulty enforcing regulations among competing nations. The increased fishing effort during this period and the lack of enforceability of regulations finally led to the ecologic collapse of fish communities in the early-1970s. Following the collapse, individual country quotas were set, and then lowered annually as the pressure on the fish continued. It also became clear at this time that the MSY formula for setting Total Allowable Catch (TAC), even if it was enforceable, was doing damage to the fish supply. A new criterion called Fo.1 was adopted which lowered the TAC by one-third. The Fo.1 formula is represented on a "catch and effort" graph as the point at which the increase in catch for a given increase in fishing effort is only 10% of the increase in catch that would occur in a virgin fishery. Canada adopted this model when it took over control of the fishing grounds in 1977.

Although there was a general sense of optimism in the period following the declaration of the 200-mile limit, fish communities were vulnerable from overexploitation, and the government had not as yet come up with any systematic way to control the expansion of the Canadian fleet. There was also the problem of dividing up the TAC among the competing interests in a rapidly changing fleet. The federal government set up a committee to allocate the fish in a way that caused the least amount of disruption to the offshore-based processing industry, and to keep to a minimum any conflicts between local and mobile fleets. The resulting 1977 groundfish plan was a complex combination of quotas based on vessel and gear type, trip limits, catch limits, and area closures that attempted to satisfy the offshore companies located in the various provinces of the East Coast.[26]

But at the same time as trying to institute conservation measures, the Canadian government — through such agencies as the Department of Regional Economic Expansion (DREE) — was pouring large amounts of money into the development of a larger Canadian fishing industry. Reflecting this gold rush mentality, the Kirby Report states:

> Provinces with no trawler fleets wanted them; provinces with trawlers wanted to add more and bigger vessels. Companies poised themselves for the growth in resources. Processing plants expanded; new ones were built. Fishermen who had left the industry since 1968 came back again; by 1980 fishermen were as numerous as they had been before. Banks loaned money with less than normal prudence. . . . While the Department of Fisheries and Oceans was slowly tightening up the licensing regime with one hand (and preaching restraint), it was passing out subsidies for fishing vessel construction with the other, as were provincial loan boards.[27]

So although the Canadian government did not issue new trawler licenses, it did provide subsidies for building larger and more efficient vessels than were previously licensed. In the small boat sector, excess capacity was addressed by dividing the fleet into full-time and part-time fishermen, with only the full-time licenses being transferable to other fishermen. This was an attempt to gradually reduce the number of inshore fishermen who had access to the resource in hopes of raising the income of those who remained in the industry.

In order to better administer the growing complexity of fishery regulations, the Department of Fisheries and Oceans was created in 1979 by the Government Organizations Act. It brought together elements that had functioned as the Fisheries and Marine Services in the previously titled Department of Fisheries and the Environment. What had previously been an organizational responsibility concerned mainly with conservation, and based on biological data, had now to encompass broader social and economic responsibilities. Comprehensive regulation required this integrated institution which also became the basis for creating the new government department.

The groundfish management plans of 1977-1981 were basically the same as the 1977 plan, aside from the larger concessions to the expanding inshore and midshore fleet. The government put in place a domestic advisory body responsible for stock assessment which replaced the international one which had advised the ICNAF, which had in turn been replaced by North Atlantic Fisheries Organization (NAFO). Unlike the international committee, the Canadian complement was composed of scientific members from the government and had no representation from the industry sector. This insured the separation of the "public interest" from the private competition which had limited the effectiveness of conservation in the international context. At the same time, the government attempted to widen the consultative process by holding seminars throughout the Maritimes so that the various sectors could make their views known. The Canadian government also helped these sectors to set up organizations that would allow them better opportunities for being heard in the policy process.

Because biotic communities were in poor shape when the government took them over in 1977, there was little room for the expansion of the Canadian fleet. Despite this reality, and arising from government programs which promoted the expansion of the fleet as well as the processing infrastructure, there was a rush to replace the recently departed foreigners. As a result, this new Canadian fleet of inshore and midshore draggers continued the pressure on vulnerable fish communities. This new fleet used the dormant dragger licenses that many of the inshore fishermen had at the time but rarely used, mainly because they were accustomed to participating in a variety of fisheries throughout the year. Furthermore, setting up for dragging was expensive and usually required a larger boat than most of them owned. These licenses were valid for vessels under 45 feet, so new boats were built to 44 feet, 11 inches and 15-20 feet wide. They had fish holds that could carry up to 50,000 pounds of fish and their bulk enabled them to travel greater distances in search of fish.

Despite the fact that sufficient catching capacity existed in the Canadian fleet in 1977, and despite the fact that the government issued no new dragger licenses, and despite the fact that the fish were vulnerable to overexploitation, the domestic fleet expanded rapidly during the late-1970s and early-1980s. This overexpansion was made more urgent because of the increased capital costs of this expansion and the higher fuel costs. In a 1980 report to the government, it was predicted that by 1985 a trawler would need to catch 90% more fish to be economically viable.[28] This increase in fishing effort could not be reconciled with conservation measures related to the projected size of the TAC for 1985.

Even though they were aware of this impending collision between conservation and development initiatives, the federal and provincial governments continued to provide subsidies and loans for the construction of these draggers. Part of the rationale behind this was an attempt to create a fleet sector that was not vertically-integrated into a large company, thereby creating a profit centre within the fishing fleet itself. Since 1977, Federal Fisheries Minister Romeo LeBlanc had been trying to achieve this split but was met with resistance from the big companies and some unions, who feared reduced bargaining power in a more decentralized market. Also, it was hoped by the government that the increased competition would raise fish prices at the wharf, which many felt were kept artificially low by the large companies who bought from their own boats.

So, in a period when the federal government was supposed to be setting in place the organizational infrastructure to manage this newly Canadianized industry, as well as increasing the profitability of those involved, the forces that had caused ecologic collapse and financial problems in the mid-1970s were still at work in the industry. Despite the mandate to bring economic development in the industry under control for "the best interests (economic and social) of Canadian society," the Canadian government merely internalized, within a national framework, the technological and economic processes which had depleted biotic communities in the international context.

As I stated earlier, one of the difficulties of implementing conservation measures in the aftermath of ecologic collapse is that the forces of development are already in place and resist the lowering of exploitation levels. By declaring the 200-mile limit and banishing the foreigners, the

Canadian government had a rare opportunity where there was an actual and immediate decrease in the forces of development which would allow for the implementation of lower exploitation levels. By so wholeheartedly supporting the expansion of the Canadian fleet, it nullified this immediate decrease in pressure. This squandered opportunity adds an element of tragedy to the situation which evokes the modern human enthrallment and absorption in the forces of development. The crisis in the early-1980s which followed this expansion of the forces of development only intensified that enthrallment.

NAVIGATING TROUBLED WATERS:
THE KIRBY TASK FORCE (1983)

By 1981 the Atlantic Canadian fishing industry was again in a serious financial crisis. The problems centred mainly on the large processing companies rather than on the fleet sector, and had much to do with the global conditions of stagflation that had developed in the late-1970s. This led to declining markets in the United States, increased competition on foreign markets because of fluctuating exchange rates from Iceland and Norway, and increased fuel and interest costs; and resulted in the processors having to carry an extremely large inventory at the same time as having to handle an increased debt load on a shrinking cash flow. In an attempt to solve these problems and address the long-term prospects of the industry, the Canadian government set up a Task Force headed by Michael Kirby in early 1982.

After an intensive consultative process that ranged across Eastern Canada, as well as being armed with its own team of researchers, the Task Force released its report in February 1983. It set out clear objectives by which it would judge the long-term aims of the industry:

> (1) The Atlantic fishing industry should be economically viable on an ongoing basis, where to be viable implies an ability to survive down-turns with only a nor-

mal business failure rate and without government assistance.

(2) Employment in the Atlantic fishing industry should be maximized subject to the constraint that those employed receive a reasonable income as a result of fishery-related activities, including fishery-related income transfer payments.

(3) Fish within the 200-mile Canadian zone should be harvested and processed by Canadians in firms owned by Canadians wherever this is consistent with Objectives 1 and 2 and with Canada's international treaty obligations.[29]

The report presents what it perceives as a significant shift in government policy:

The order of priority given to Objectives 1 and 2 is of fundamental significance because, if adopted by the federal government, it would be almost universally regarded as a definite change in government policy. Until now, it has appeared to most people familiar with the Atlantic fishery that federal government decisions have been based on Objective 1 and 2 being in reverse order of priority or, at most, equal in priority.[30]

Conscious of earlier government policy with regard to the fishery, the report continues:

The solution to these problems cannot be found in a static society or in a static fishing industry, forever dependent on the taxpayer for supplements to bring its meager earnings up to subsistence.[31]

So, if in 1977 the Federal government gained full jurisdictional responsibility for the fish and expanded its role in conservation, in 1982 it went beyond its previous ad hoc policies with regard to industrial development in an effort to establish the industry on a sound financial footing. But in contrast to the larger role it played in conservation, the government wanted to get away from ongoing economic bailouts and subsidies, and leave the running of the industry to a healthy private sector. This attempt to set out a comprehensive fisheries policy in economic terms was expressed in a brief presented to the Kirby Task Force by the Fisheries Council of Canada:

> It is essential that the Government of Canada establish an explicit, overall objective for the Canadian fisheries and that its Department of Fisheries and Oceans articulate that objective in terms sufficiently specific to allow both government and industry to plan and implement operational policies.[32]

The federal government accepted virtually all the Task Force's 57 recommendations. These recommendations included ways to improve marketing strategy and fish quality, increase processing plant efficiency, and stabilize fishermen's income through unemployment insurance payments, and proposals for including foreign trawlers in the fisheries management plan.

Many of the problems addressed by the Kirby Report were effects of the rapid expansion of the Canadian industry following the declaration of the 200-mile limit. But what was once again made clear, as had been the case in earlier reports, was that the government perceived the common property aspect of fishery as being at the root of the problem of over-expansion:

> It must be further acknowledged that the fishery is rife with incentives to expand. The most important of these are the common property nature of the resource.[33]

Although the government had attempted to limit the expansion of catching capacity through limited-entry licensing and restrictions on increases in the size and catching capacity of vessels, none of these measures could curtail the powerful logic of expansion in what the report saw as the inherent problem:

the exploitation of a common property resource. It is this whirlpool of chronic inefficiency and instability that P. H. Pearse addressed with regard to Canada's West Coast fishery. However, its conclusions concerning how the problems related to the exploitation of common property resources are perceived by fishery managers are pervasive:

> The central problem of the commercial fisheries is the chronic overcapacity of the fleets. . . . This wasteful pattern of development reflects governments' failure, in spite of repeated attempts, to develop a policy that would encourage the industry to develop efficiently. . . . In recent years, licensing systems in considerable variety have been designed to alleviate the problem, though few can be said to have had much beneficial effect. . . . The greatest single challenge is reorganizing the policy framework for the commercial fisheries to stop this treadmill of overcapacity, and further to reduce the present excess capacity, so that fishermen can receive reasonable returns and the people of Canada can begin to realize some of the substantial surplus that the fisheries are capable of yielding with a better fleet structure.[34]

The Kirby Report goes on to state that "the fishery is the only common property resource in Canada that is not allocated, at least in part, by a quasi-judicial process. The common property of the airwaves, for instance, is regulated by the CRTC."[35] Recognizing that limited-entry licensing, as well as gear and vessel restrictions had not worked, the Kirby Report recommended a system of quota licenses which had also been recommended by Pearse for the West Coast industry. This program would

> . . . allocate annual catch quotas. . . . The quotas would normally be expressed as a percentage of the Total Allowable Catch for a given stock. They could

be transferred or sold among fishermen subject to restrictions the government might wish to impose. . . . Once a fisherman had a guaranteed quota, he could then plan to catch it at the least cost. . . . In essence, the quota licence gives a fisherman an individual property right to a certain amount of the previously common property resource. It thus 'privatizes' the common property.[36]

This granting of private property rights was expanded throughout the 1980s. In its recommendations, the Task Force Report concludes:

The Task Force attached great importance to the need to come to grips with the problems caused by the common property nature of the fishery. We believe that the resolution of these problems will be of enormous benefit to fishermen.[37]

As a result of the groundfish crisis of 1982, which was concentrated in the financial problems of the large vertically-integrated companies, DFO instituted an enterprise allocation program (vessels greater than 100 feet) for these offshore companies in an effort to make them more financially stable. The rest of the fishery remained on a limited entry, trip quota system. Because the majority of the overcapacity (in terms of percentage of TAC) existed in the inshore dragger fleet, this did little to solve the overcapacity problem in the fishery as a whole. What the Kirby Task Force did do is focus fishery policy on large-scale, highly-industrialized priorities, and in doing so, initiated a kind of watershed whereby the common property nature of the artisanal fishery was marginalized in favour of modern rational business practice. This transformation had significant consequences with regard to how the relationship between conservation and development was understood. Central to this transformation — and to free trade and globalization generally — is the privatization of property rights in the hands of industrial interests at the expense of local, artisanal communities.

Trying to find some strengths in the fishery and some reason for optimism, the report states that:

> The state of the Atlantic fishery in 1982 is not a story of unalloyed disaster. There are two very significant bright spots: the outlook for the harvest and the vast unexploited potential for improved quality and cost efficiency in the industry. Finding ways to capture this potential is a principal theme of the rest of this Report. Our comments on the resource will be almost perfunctory, because we believe its management is quite well in hand.[38]

A consummation devoutly to be wished. These two supposed bright spots dimmed considerably as programs such as dock-side grading, which was supposed to improve quality, bogged down in procedural wrangling between government and industry and were therefore never implemented. Furthermore, the efficiency of the fleet caused ecologic collapse in spite of the capable hands of the Canadian government. Rather than facing "the happy prospect of a 50% increase in the groundfish catch between 1981 and 1987,"[39] signs of another downturn were imminent by the mid-1980s, and by 1989 a new and more severe crisis was at hand.

SCOTIA-FUNDY GROUNDFISH TASK FORCE (1989)

Throughout the mid-1980s, the inshore draggers (45-64 feet) continued to press for a larger share of the quota to relieve the overcapacity. Rampant misreporting of catches and antagonism between the inshore fleet and DFO were the order of the day as prices rose in the context of an increasingly scarce commodity. This led to increased pressure at the moment of sharp decline in landings. Because the inshore dragger fleet was concentrated in the Scotia-Fundy region, there had been excess pressure on the fish. By June of 1989, the fleet had caught the whole quota for the 1989 calendar year. The cod and pollock were seriously depleted and there was a threat of a total collapse of the haddock.

The intense pressure put on biotic communities in the midst of collapse is part of a repeated phenomenon affecting a wide range of exploitation patterns. In the *State of the World 1995*, Lester Brown states:

> When sustainable yield thresholds are crossed, the traditional responses proposed by economists no longer work. One common reaction to scarcity, for instance, is to invest more in production. Thus the key to alleviating seafood scarcity is to invest more in fishing trawlers. But in today's world this only exacerbates the scarcity, hastening the collapse of the fishery. Similarly, as food prices rise, there is a temptation to spend more on irrigation. But where water tables are already falling, investing in more wells simply accelerates the depletion of the aquifer and the eventual decline of irrigation.[40]

This is the reality which drove the expansion of the fleet in the mid-1980s. As well as there being an increase of exploitation due to scarcity, prices also rose because fish became the food of choice (for health reasons) during this period.

The resulting overcapacity in the groundfish fleet resulted in the setting up of the Scotia-Fundy Groundfish Task Force, chaired by Jean Haché. After a series of hearings throughout the region, the Task Force findings were published in December of 1989. As in previous inquiries into crises in the industry, the goal of the Task Force was to "develop recommendations leading to. . . long-term stability and prosperity in the groundfish industry."[41] While reiterating the three main objectives of the Kirby Task Force — 1) economic viability of the industry; 2) maximization of employment at reasonable incomes, and 3) the Canadianization of the industry — the Haché Report conceded that these goals could not be achieved as long as the industry was based on a common property resource:

> The emphasis on economic viability, income and employment reflects an evolution in the philosophy of fisheries management from an exclusive focus on conservation to a concern about the economic and social impact of fisheries management decisions. This shift has not undermined the importance of conservation, but rather indicates an increased appreciation that conservation alone is not enough. The familiar "race to fish" leads to too much investment in fishing capacity, which puts the fishing industry in a fragile financial state even when fishing is controlled well enough to protect the stocks.[42]

The present system, the report concedes, has "imposed an unnecessary strain on both Departmental enforcement resources and on fishermen."[43] This report has to be seen within the broad background of a Progressive-Conservative Canadian government whose main policy initiatives emphasize privatization, deregulation, and free trade, all in the name of economic globalization. Therefore, the phrase "focus on conservation" refers to the comprehensive regulatory framework set up on the eve of the declaration of the 200-mile limit which had become anathema to this governmental perspective linked to globalization. Preoccupation with government spending on these types of programs came under attack, especially when it was perceived that fishery regulation was a very expensive failure anyway. All the conventional wisdom of the day argued for turning the industry over to market forces through granting ownership of the fish to the participants in the industry.

With the expanded role of government in fisheries regulation, the earlier argument that all Canadians should have an opportunity to exploit the nationalized fishing grounds is replaced by a concern that Canadian taxpayers should not be saddled with paying for a complex set of regulations and subsidies that regulates an unstable and inefficient industry, and that only benefits a limited number of licensed participants. To quote the Haché Report:

> Fisheries management employs public resources to generate private gain. The process should be made as efficient as possible to minimize the cost to Canadian taxpayers. Management has evolved toward a system demanding a high degree of administrative, scientific, and enforcement support while manpower and financial resources have been declining. In this light more efficient management measures must be sought. In addition, the management regime should be designed to encourage fishermen to become more efficient, which should in turn produce more benefits overall.[44]

After listing the failures of DFO to limit catching effort through licensing policy, trip quotas, area closures, and gear and vessel restrictions, the Haché Report states that "unless the basic management structure is changed, the cycle is doomed to repeat itself."[45] To replace the old structure, the report proposes a new classification of the fishery sectors. Vessels over 65 feet (instituted in 1988) and vessels over 100 feet (instituted in 1982) would continue with their enterprise allocation program which gives them access to a prescribed share of the offshore quota. This granting of a share of tradable quota — in effect, privatizing a share of the quota as a transferable piece of property — was perceived to minimize conflict between regulators and exploiters, at the same time as promoting economic efficiency. The inshore and nearshore fleets would be divided into three groups:

> Group A- A large proportion of all groundfish licenses are issued for vessels engaged in the inshore mixed fishery. Groundfish is used to supplement fishing for lobster, scallops, herring, mackerel and other species. Group A vessels are limited to a six month season and 1,500 kg per trip.

Group A also is restricted to fixed gear (longline, gillnet) and would usually involve boats in the 35 foot range.

Group B- The Task Force concluded that the overcapacity problems in the inshore fixed gear sectors are less critical at this time than in the mobile gear fleets. Group B, then, consists primarily of fixed gear vessels whose catching capacity is higher than the 1,500 kg allocated for Group A. This group is subject to an overall quota, similar to what is in place now and as soon as that annual quota is caught, the fishery is closed. These vessels can also retain licenses for other fisheries.

Group C- This group consists of vessels that represent the most critical overcapacity problem in the inshore fleet, and is limited to vessels under 65 feet with mobile(dragger) gear licenses. Competition for fish has generated this overcapacity. "A fundamental change in incentives is required." For 1990, Group C will continue to fish competitively with season and trip limits. When required legislation, administrative arrangements and monitoring systems are in place, Group C licence holders will jointly choose whether to move to:

> 1) individual vessel quotas with pooling or partner ship provisions
>
> 2) individual transferable quotas
>
> 3) an ongoing arrangement for self-funded retirement of licenses
>
> 4) a competitive quota system as at present
>
> 5) other options acceptable to the industry and DFO.[46]

The licenses for group C would cost thousands of dollars annually as it entitles the fisherman to a certain share of the fish. This money would also offset the increased monitoring costs, as well as funding a buy-back of licenses of those who want to leave the industry.

Thus, in order to increase the profitability of the industry and make the government regulatory agency more effective, the Haché Report moved toward ending exploitation patterns which it attributed to common property. By doing this, it hoped to resolve the inherent conflict between what it saw as the expansionist logic of common property economic dynamics originally outlined by Gordon, and the need for catch limits to preserve biotic communities. Despairing of managing exploitation through a system of comprehensive regulation — which was perceived to have created both large government expenditure and industry inefficiency — the Haché Report acceded to the logic of a market economy and expanded the individual ownership of the Total Allowable Catch of fish, in hopes that this new approach will increase self-regulation and promote economic stability and ecologic viability.

As will be discussed in Chapter IV, this blaming of common property for the ills in the fishery misconstrues the causes of industrial overcapacity and ecological destruction. It is rational behaviour in a market economy which causes these problems. The difficulty that regulators have in limiting exploitation has also to do with rational behaviour in a market economy. By withdrawing its mandate as the "unified directing power" in the fishery — as it was set out in 1976 — the Canadian government recognized that comprehensive regulation had failed, but it misunderstood the reasons for the failure. Because it misunderstood the problem, it turned the fishery over to the very forces of modern market economy which were in fact destroying it.

CHARTING A NEW COURSE: TOWARD THE FISHERY OF THE FUTURE, THE CASHIN TASK FORCE REPORT (1993)

Failure of the [groundfish] resource means a calamity that threatens the existence of many of these communities throughout Canada's Atlantic coast, and the collapse of a whole society. . . . We are dealing here with a famine of biblical scale - a great destruction.

Richard Cashin[47]

On July 2, 1992, Canadian Minister of Fisheries John Crosbie declared a moratorium on fishing for northern cod in a range of areas off Atlantic Canada for a two year period. Since that original declaration, area closures have expanded and the moratorium's time frame — originally set at two years — has been extended into the next century.

In the aftermath of ecologic collapse, the Canadian government set up the *Task Force on Incomes and Adjustment in the Atlantic Fishery* chaired by Richard Cashin, to look into the causes of the collapse and what it meant for those people living in Atlantic coastal communities who depended in the fishery. As the report states:

> . . . we have tried to describe the Atlantic fishery, to set out its problems, and to recommend ways to break the cycle of overdependence, excessive pressure on a finite resource that is the fish stock and overcapacity in both harvesting and processing, ultimately resulting in chronically low and unstable incomes.[48]

In terms of setting out the problem, the report remains circumspect in pointing to the cause of the "great destruction":

> The reasons for the collapse are complex, and not well understood — but the consequences are all too clear: devastation for those who live by the groundfish. Groundfish have always been subject to cyclical swings. There are ecological changes and anomalies that effect their reproduction and survival, such as changes in water temperature and salinity. Changes in their food supply can have a major impact, especially on cod. The impact of predators such as seals is also a major factor.

> When a forage stock such as capelin declines, cod will prey
> on less nutritious food, including small cod.[49]

These comments are made before there is any discussion of human exploitation of the cod, and leaves the impression that the collapse was caused by the cod eating their young. In the following paragraph, the report states that, in 1990, there were 28,000 boats registered, employing 64,000 people in Atlantic Canada, along with 800 fish plants employing 60,000 people, with no mention that this activity might have a detrimental effect on fish populations. Rather, it was meant to emphasize the significance of the catastrophe in human terms.

When the Cashin Report does discuss human exploitation of the natural world, it presents the problem in terms of the "overdependence on the fishery, pressure on the resource, and industry overcapacity — all interacting in a vicious cycle."[50] Overdependence means that "there are more people and capacity than the fishery can maintain." It is caused by "a social tradition of the right to fish among Atlantic Canadians," by a "lack of economic alternatives," as well as "the use of the fishery as the employer of last resort." Pressure on the resource means that:

> . . . while the resource is finite, human population keeps
> growing. Ignorance is another factor. This includes our lack
> of adequate knowledge of the resource, its habitat, the inter-
> action among species, and other ecological factors. In addi-
> tion, fishing technology keeps improving. Pressure also comes
> about from mismanagement of the resource — the failure to
> control, to enforce limits, and the lack of a meaningful part-
> nership with the users of the resource. And it comes from
> wasteful harvesting practices.[51]

The third aspect of this vicious cycle is overcapacity and results in a situation where:

Too many harvesters use too many boats with too much gear trying to supply too many processing plants by finding and catching too few fish. The results are low and unstable incomes, problems with income assistance, especially unemployment insurance, and a generally unprofitable industry, characterized by persistently underfinanced operations. The net effect exacerbates the problems of overdependence and pressure on the resource. And the cycle continues.[52]

The report then states that the chief cause of this cycle is the common property nature of exploitation in which "everyone wants as much as the resource as possible."[53] This argument links this report with government inquiries which had come before it and which had identified common property as the problem in the fishery.

As I argued in the introduction, I regard the fishery as a significant case study of the intellectual ruin in which we live, called economic development. The Northwest Atlantic fishery exhibits components of overexploitation which are entirely typical of environmental problems all over the world. This is what makes it such a good case study for the practice of conservation which can make a contribution to the theory of sustainability. All of this discussion is framed within historical realities related to the imperatives of modern economic and technological priorities, and would be undiscussable without them. But these government reports on the problems in the fishery make no attempt to recognize that fishery problems are occurring at the same time as a wide range of other environmental problems, and that these issues are not just fishery issues but recur in a wide range of modern human activities. In order to make this linkage, these reports would have to begin to question the imperatives of modern market economy, something they will not do.

There is an element of truth to all the aspects of fishery problems that the Cashin Report identifies, but they mean nothing without a significant recognition of the pressure modern economy exerts on the natural world. It is this reality — combined with the general failure of national regulatory frameworks — which informs and underwrites "the fundamental problems in the fishery," as identified by the Cashin Report. It is the refusal to challenge the forces of modern economy which condemns conservation strategies to failure. The Haché Report states that the "focus on conservation" caused "unnecessary strain" on fishery managers. This rationale was used to expand private property in the fishery and turn over management to the larger players in the industry. The conflict between conservation — as conceived of as a national public regulatory mandate — and development was seen to cause needless strife and inefficiency, as well as being a general tax burden to Canadians. But national regulatory frameworks failed not because they put restrictions in the way of development and caused undue conflict. They failed because they did not adequately restrict the forces of development. Conservation in the fishery failed because it was compromised and implicated at every turn by the forces of development, and did not make it sufficiently problematic. Attempting to operate within the scientific, technological, and economic realities of modern economic development, conservation initiatives were the captive of the forces that were destroying biotic communities in the Northwest Atlantic.

The Cashin Report states that a management perspective which gives

> priority to conservation, even when stocks can be accurately measured, is not by itself effective management of the resource if other concerns are disregarded. . . [This is reflected in] the failure of the Department of Fisheries and Oceans to manage the fishery as a whole, considering all its aspects, and the range of social and economic impacts of management decisions.[54]

Once again, the focus on conservation is seen as causing the problems in the fishery, such as overcapacity and inefficiency. And once again, although it identifies a trouble spot, the report misconstrues the reasons for it because it will not make modern economy problematic. I argued in the introduction that conservation is not an on/off switch for destructive behaviour imposed by an external authority at some upper level of exploitation at the last minute. This version of conservation is reflected in the quota system as practiced in the fishery. Conservation is doomed to failure if all the modern economic and technological realities are allowed to be present in the exploitation of the natural world, and then are suddenly expected to cease operating when a specified catch level is reached. This is a profoundly unrealistic expectation, and is the reason why conservation has failed in the Northwest Atlantic. Historical forces do not respond to someone blowing a whistle. It may have enticed the mice out of Ireland, but it will not stop the expansion of capital.

The granting of private property rights through enterprise allocation (EAs) and individual transferable quotas (ITQs) to the larger interests in the industry were supposed to rid the fishery of the inefficiency of whistle blowing fishery officers, and promote a more rational use of the resource because competition for it under the quota system was circumvented. Contrary to this point of view, the *Report of the Workshop on Scotia-Fundy Groundfish Management from 1977 to 1993* published by DFO, states:

> . . . all of the analyses [of fish catches] inferred dumping, discarding, and highgrading (by both mobile gear and fixed gear) due to trip limits, EAs, ITQs, and imbalances between quota and abundance for the CHP species mix in a given area. The observer evidence infers that these practices have increased rather than decreased within introduction of property at the level of an enterprise. The port technician anecdotal information and the interviews infers that property at the individual level (IQs) is providing more incentives for illegal fishing practices at sea, but the levels of such practices cannot be quantified.[55]

What this statement makes clear is that "inefficient" exploitation patterns are not mitigated by the granting of property rights. The statement quoted above that property rights are "providing more incentive for illegal fishing practices at sea" is made in the body of the Groundfish Management Report 1977-1993. The introduction to the report sets out the standard view that the "race to fish" inherent to common property has caused many of the problems in the fishery, as well as causing undue strife between regulators and exploiters. The Department of Fisheries and Oceans therefore embarked on a program of privatizing fish as a way of integrating conservation into development:

> These programs aimed to change the fundamental motivations in common property systems by issuing quotas to companies or individual fishers. Enhanced ownership is expected to mitigate the "race for quota" and allow individual fishers to tailor capacity and ultimately to target fishing effort to the quotas they control. Fishers would then be able to maximize profits without having to maximize the volume for competitive quotas. EA programs were introduced to offshore fleets in 1982, vessels 65-100 ft. in 1988, and to the inshore mobile gear fleet in 1991.[56]

In keeping with the Canadian Government reports on the ongoing crises in the fishery that had come before it, there is once again evidence of a profound analytical failure in the aftermath of ecologic collapse, and a resistance to recognizing that standard practice in a modern economy caused the "great destruction" in the Northwest Atlantic. As Wolfgang Sachs states with regard to this kind of resistance to environmental problems:

> Reaffirming the centrality of 'development' in the international discussion on the environment surely helps to secure the collaboration of the dominating actors in government, economy, and science, but it prevents the rupture required to head off the multifaceted dangers for the future of mankind. It locks the perception of the ecological predicament into the very world view which stimulates the pernicious dynamics, and hands the action over

to those social forces — governments, agencies, corporations — which have largely been responsible for the present state of affairs.[57]

The "present state of affairs" is general woe in the Northwest Atlantic. Despite the catastrophe which has befallen the fishery, those responsible for it remain in charge, and resist the recognition that they are rattling around inside an intellectual ruin called development. *The Canadian Maritimes Fishery: Let's Fix It* — an action plan put together by fishers in the Southwest Nova Fixed Gear Association — conveys the issue this way:

> Despite their admitted mismanagement, the 'managers' continue to defend their capability to reform their own department, the management process, and the industry. Those of us who have endured the pain of their past mistakes have little faith that they will resolve problems they haven't grasped.[58]

Instead, those in power engage in demonizing the Spanish foreigner in high seas conflict and confrontation, all the while making high-minded comments about the importance of conservation. In the aftermath of collapse, regulation of the fishery occurs at the end of a gun barrel pointed outwards, rather than analytical questions aimed inward.

NOTES

1   Bruce Mitchell. 1989. *Geography and Resource Analysis*. London: Longman & Wiley, p. 3.

2   Arthur McEvoy. 1987. "Toward an Interactive Theory of Nature and Culture: Ecology, Production, and Cognition in the California Fishing Industry." *Environmental Review*. Vol. 11, No. 4, p. 295.

3   *Law of the Sea Discussion Paper*. 1974. Ottawa: Department of External Affairs, p. 3.

4   Fisheries and Marine Service. 1976. *Policy for Canada's Commercial Fisheries*. Ottawa: Department of the Environment, p. 63-64.

5   IUCN, UNEP & WWF. 1980. *World Conservation Strategy*. Gland, Section 7.

6   H. Scott Gordon. 1954. "The Economic Theory of the Common Property Resource: The Fishery." *Journal of Political Economy*, Vol. 62, p. 124.

7   Gordon. (1954:125).

8   Gordon. (1954:128).

9   Gordon. (1954:128).

10 Gordon. (1954:129).

11 Gordon. (1954:135).

12 Parzival Copes. 1977. "Canada's Atlantic Coast Fisheries: Policy Development and the Impact of Extended Jurisdiction." Burnaby: Simon Fraser University, p. 19.

13 *Policy For Canada's Commercial Fisheries*. (1976:47).

14 PCCF. (1976:49-50).

15 PCCF. (1976:53).

16 PCCF. (1976:53).

17 PCCF. (1976:50).

18 PCCF. (1976:39).

19 PCCF. (1976:41).

20 PCCF. (1976:20).

21 PCCF. (1976:50).

22 PCCF. (1976:50).

23 PCCF. (1976:58).

24 PCCF. (1976:61).

25 PCCF. (1976:63).

26 R. D. S. Macdonald. 1984. "Canadian Fisheries Policy and the Development of Atlantic Coast Groundfisheries Management." *Atlantic Fisheries and Coastal Communities: Fisheries Decision-Making Case Studies*. [

Cynthia Lamson and Arthur J. Hanson eds.] Halifax: Dalhousie Ocean Studies programme. p. 43.

27   M. J. L. Kirby. 1983. *Navigating Troubled Waters: Report for the Task Force on the Atlantic Fisheries*. Ottawa: Minister of Supply and Services, p. 20.

28   P. R. Hood, R. D. S. Macdonald, & G. Carpentier. 1980. "Atlantic Coast Groundfish Trawler Study." Ottawa: Department of Fisheries and Oceans.

29   Kirby. (1983:186).

30   Kirby .(1983:12).

31   Kirby. (1983:8).

32   Kirby. (1983:185).

33   Kirby. (1983:32).

34   P. H. Pearse. 1982. *Turning the Tide: A New Policy for Canada's Pacific Fisheries*. Ottawa: Governor General of Canada, p. 75-76.

35   Kirby. (1983:214).

36   Kirby. (1983:218-219).

37   Kirby. (1983:223).

38   Kirby. (1983:23).

39   Kirby. (1983:23).

40   Lester Brown. 1995. *State of the World 1995*. New York: Norton, p. 15.

41   Jean Haché. 1989. *Report of the Scotia-Fundy Groundfish Task Force*. Ottawa: Ministry of Supply and Services, p. 7.

42   Haché. (1989:9).

43   Haché. (1989:10).

44   Haché. (1989:10).

45   Haché. (1989:23).

46   Haché. (1989:53-55).

47   Richard Cashin. 1993. *Charting a New Course: Toward the Fishery of the Future*. Ottawa: Minister of Supply and Services, p. vi-vii.

48   Cashin. (1993:v).

49   Cashin. (1993:5-6).

50   Cashin. (1993:14).

51   Cashin. (1993:14).

52   Cashin. (1993:14).

53   Cashin. (1993:15).

54   Cashin. (1993:17).

55   J. R. Angel, R. N. O'Boyle, F. G. Peacock, M. Sinclair, & K. C. T. Zwanenburg. 1994. *Report of the Workshop on Scotia-Fundy Groundfish Management from 1977 to 1993*. Can. Tech. Rep. Aquat. Sci. 1979: vi, p:115.

56   Angel. (1994:2).

57   Wolfgang Sachs. 1993. "Global Ecology and the Shadow of 'Development'" *Global Ecology*. London: Zed Books, p. 3-4.

58   Southwest Nova Fixed Gear Association. 1995. *The Canadian Maritimes Fishery: Let's Fix It*, p. 13.

IV

Resource Analysis:
Fishery Science in the Northwest Atlantic

. . . resource analys[is] seeks to understand the fundamental characteristics of natural resources and the processes through which they are, could be, or should be allocated and utilized.

Bruce Mitchell1 [1]

. . . I'm willing to go so far as to say the forecasting tools at the foundation of the scientific-rational basis of industrial society are all biased in the same dangerous way. . . . And none of us are seeing the problem, let alone dealing with it.

Jake Rice [2]

I n an interim report for an inquiry entitled the *Independent Review of the State of the Northern Cod Stock* (1990) which was prepared for the Minister for the Federal Department of Fisheries and Oceans, Leslie Harris describes the short-term goals of his panel this way:

. . . the Panel focused its attention upon the suitability of the mathematical modeling techniques employed by DFO scientists, a preliminary examination of the quality of the data inputs into the model, an assessment of the appropriateness of the management advice that has been offered to the Minister, and the identification of some interim measures that might assist in improving the reliability of advice for 1990 and beyond. [3]

Along with the many other social and economic problems that have plagued the Northwest Atlantic fishery, answering the very basic but difficult question about how many fish there are to catch, always remained an unknown factor that threatened the viability of the industry. In no uncertain terms, Harris states that "the Panel is persuaded that there has been a serious underestimate of fishing mortality rates in the years between 1977 and 1989."[4]This resulted in the Total Allowable Catch being set at double what it should have been and necessitated, as Harris argued in 1989, large cutbacks in order to insure the survival of the northern cod. Although Harris' advice was resisted at the time by the Minister of Fisheries and Oceans, the state of the cod turned out to be even worse than the Harris Report claimed.

The time period of 1977 to 1989 in which Harris claimed there was serious overfishing, corresponds to the whole of Canada's mandate as manager of the fishery. This failure to correctly assess the size of biotic communities is in direct contradiction to the Canadian Government's avowed mandate in declaring the 200-mile limit, as presented in *Policy For Canada's Commercial Fisheries* (1976):

> Institute a co-ordinated research and administrative capability to control fishery resource use on an ecological basis and in accordance with the best interest (economic and social) of Canadian society.[5]

In his book *Fishing For Truth: A Sociological Analysis of Northern Cod Stock Assessments from 1970 to 1990*, A. C. Finlayson presents the problem identified by the Harris Report in the way:

> In the current atmosphere of social, economic, and environmental crisis, everyone with an interest in the fishery is searching for the reason for this latest failure. Many fingers are being pointed at traditional targets from previous crises. Among these are: overfishing (both domestic and foreign), federal mismanagement for reasons of political expediency,

and over-capacity in the harvesting and processing sectors. But in the latest crisis, voices in all sectors of the fishing industry — the federal management structure, the media, and the general public — suggest that, even given other (contributing) factors, science, the erstwhile "saviour," is not the solution but part of the problem.[6]

This perception of fishery science as "part of the problem" runs counter to the perspective which underwrote Canada's claim to the Northwest Atlantic. As Finlayson states:

> A broadly-shared, quantified reality, the rights of sovereign states, and Canada's unsurpassed expertise in fishery science were overwhelming rhetorical as well as organizational forces. . . . Canadian scientists believed that the theory of fish population dynamics was reasonably well understood. What had prevented rational, sustainable management in the past had been lack of authority, control, and resources. Now they had all three.[7]

It is this same kind of claim to "authority, control, and resources" which underwrites so many current sustainability strategies concerned with environmental problems. This claim relies on a series of assumptions about the natural world:

> (1) The universe is mechanistic and deterministic and its workings are governed by a few fundamental and unvarying Laws.
> (2) The marine ecosystem and its sub-systems (in this case, commercially valuable fish stocks) are fundamentally robust. That is, they are relatively insensitive to small perturbations and tend to vary around natural dynamic equilibrium states.
> (3) These natural equilibrium states are dominated (or can be

described and represented) by relatively few significant variables. In this case, they are fecundity, recruitment, natural mortality, and fishing mortality.

(4) These variables are knowable and their effects on the stocks are simple, continuous, and can be realistically modeled by an equation with a small number of parameters. Therefore, they are predictable.

(5) Science-based management can manipulate some of these variables (primarily fishing mortality). It can monitor the other to effectively control the system and produce (within certain broad limits) equilibrium states in general harmony with human needs and desires.

(6) Having rebuilt the stocks to the desired level [following their collapse in the early 1970s], they could then be maintained at that level by relatively minor adjustments in the TACs. Long-sought-for stability could be brought to the fishery and its industries.[8]

This is the "focus on conservation" which formed the cornerstone of Canadian management of the fishery, and which existed in conjunction with expanding social and economic concerns related to management. But it is a "focus on conservation" which is dominated by a production model view of the natural world, and like management frameworks, is compromised by too readily accepting the edicts of development-based approaches to the exploitation of natural processes. This kind of knowability of nature is entirely linked to a conception of nature associated with its capability to produce an annual surplus for exploitation. It is this narrow slit of recognition related to production capability which forms the basis of stock assessment models.

To the other manifold and wondrous occurrences in nature, these constructs are entirely dumb and blind. It is this dumbness and blindness related to the modern human conception of the natural world which is at the heart of the "intellectual ruin" called economic development. But, of course, it is this same dumbness and blindness which afflicts our view of human society as well. Its health, or the lack of it, is almost entirely tied to economic growth and jobs, and we find it hard to value ourselves personally outside of these realities. Why else do parents who stay at home with their children have a lower confidence level than most groups in modern society? In order to resist inclusion in the impoverished sociality of this production perspective on the relationship between humans and the rest of nature, I do not use terms such as resource, biomass, or fish stocks to describe the beings that inhabit the ocean, although I, of course, discuss others who do. This may seem like a series of unusual points to make at the beginning of a discussion of fishery science and its role in the ecologic collapse in the Northwest Atlantic, but I believe this kind of perspective is central to understanding the fishery crisis not as a unique phenomena, but as part of a range of environmental problems related to a wider crisis of modernity.

Conservation in the fishery, as well as sustainability initiatives generally, require a prescribed level of exploitation by their very definition. It is therefore the role of those biologists involved in conservation or sustainability initiatives to supply that number. Here is the principle, and here is the interest in nature, so exploit to here and stop. This natural sink can absorb this many toxins, so pollute to here and stop. This is what I mean by a production model view of nature. But these are entirely human concerns. They are at most times entirely economic concerns, and have nothing to do with the interactions in natural communities. As Leslie Harris states with regard to these production models:

> The danger in all modeling, in my view, is that you become trapped by it to some extent. It's self-fulfilling. You're dealing with data which are manipulable and variable and uncertain. You have a variety of ways that you can interpret the data. If you've got a model that you believe in you will interpret the data in a way that makes the model work. I don't think there's any dishonesty in this, as such. . .[9]

We assume that if we know nature well enough, have good enough data, and good enough models for natural interactions, the magic number related to sustainable levels of exploitation or pollution will appear. But that number is no where to be found in nature, and is in fact a profoundly impoverished expectation which is required by the priorities of the economic balance sheet.

What will be examined in this chapter are the evolving mathematical models for assessing the size of biotic communities in the Northwest Atlantic, beginning with the Maximum Sustainable Yield production models based on the relationship between catch and fishing effort that were used by the International Commission for the Northwest Atlantic Fisheries (ICNAF). Following the declaration of the 200-mile limit, Canada expanded upon models based on the Fo.1 level of exploitation, which retained the same assumptions as production models but established more conservative catch levels. Also, Canadian fishery scientists developed forms of analysis based on the age-classes in the fish population which, in conjunction with research vessel information, gave a more detailed understanding of the complexity of the marine environment and lessened reliance on commercial catch data. Despite these improvements in stock assessment models, Harris argued that there were serious problems not only with the attempt to accurately assess fish population, but also with the economic and institutional context in which this project existed:

> Perhaps it is easier, and therefore, more tempting, to seek answers through mathematical manipulations, whereas, the true solution may only become apparent when we have a more comprehensive knowledge of the biology and behavioral characteristics of the species with which we are particularly concerned and of the ecosystem in which it functions.

But as I have argued in the introduction, when development precedes conservation (or when resource development precedes resource analysis) to the point of ecologic collapse, the already dubious project of developing a production model of natural interactions becomes even more questionable. How do you understand a forest after it has been clear-cut? This is what fishery scientists were trying to do with fish communities in the Northwest Atlantic. If you only begin to try and understand the interactions of fish after you have exploited them to the point of collapse, what are your chances? If the majority of the data you base your assessments on come from exploiters who have a general mistrust of managers and who lie about their fishing activity on a regular basis, what are your chances? If fish are invisible and can only be experienced in the context of various forms of human exploitation, what are your chances?

It is the modern economic brinkmanship with the limits of natural processes in withstanding exploitation that forms the basis of the production model of nature. It is this paranoid cliff-edge which sustainability initiatives attempt to establish. It is this cliff-edge which humans will, at best, walk for the rest of history. The more pervasive the edicts of economic development and exploitation are in a society, the more central are the views of nature in productive terms, both because of the increased importance of the mindset of development and also because natural processes are exploited to their limits and require attention if economic outputs are to be maintained.

In 1954, Scott Gordon stated that "the present state of knowledge is that a great deal is known about the biology of the various commercial species. . ." Given the recent machinations involving the accuracy of fishery science predictions, this is an interesting statement of confidence especially since it was expressed forty years ago. It becomes increasingly interesting if we accept that Gordon's statement was true at the time. In other words, what is different now is the expectations of fishery science. In the context of ecologic brinkmanship, the only important knowledge is that which guides exploitation along the paranoid cliff-edge of collapse. When this is not the primary requirement — as it was not in Gordon's time when the ocean was considered infinite — it was possible to make such a statement because this magic number wasn't required.

This kind of transformation is evident in the forces exerted on biological science in the Northwest Atlantic in the 1960s. As fishery scientist Edward Sandeman, then Director of the Science Branch for Canada, states:

> It was during this period that the focus of fisheries science changed to a mathematical approach and the modern science of fishery population dynamics really took off. This was really quite a difficult time for those in fishery science because they were neither trained nor even had an aptitude for this new discipline.
>
> Fishery scientists of that era were trained to taxonomy and the microscope, and it was a very difficult challenge to change from biology to mathematics. . . . The push didn't really develop until 1970 when most of the ICNAF community started to realize that there were problems. That gross over-fishing was taking place. . . . And I guess really that's when our scientists were forced to become much more mathematically oriented, and to use the tools of population dynamics.[10]

It is clear from Sandeman's comments that we see a corresponding transformation in the economic activities associated with the exploitation of fish, and the forms of production-oriented knowledge which accompanies it along the cliff-edge of collapse. This results in the conception of the relationship between science and economics in terms of commonality, as expressed more recently by fishery scientist J. J. Maguire: "If we want to stay in business we better get closer to the clients [commercial industry]. It's straight free-market economic forces."[11] This perspective turns fishery science — as it is transformed into management decisions — into a kind of service industry providing information about the availability of raw material to industry so that it can do long-term financial planning. As the biological perspective of "taxonomy and the microscope" was gradually captured by large-scale industry assumptions, the perspective of small-scale inshore fishing people was increasingly marginalized. To quote fishery scientist Sandeman:

For the most part the majority of them [inshore fishers] have a litany of mumbo jumbo which they bring forth each time you talk to them. About where the fish are and why they're not here. They relate it to things like the berries on the trees. . . . When I was going around trying to understand a bit more about Newfoundland and the fishery, I just got completely turned off by the inshore fishermen and their views. Because they were totally unscientific![12]

This marginalization in knowledge terms of inshore perspectives, or should I say community perspectives, mirrors the marginalization of this group in the policy process, as well as in terms of the industrial claim to the fish in the Northwest Atlantic. What is clear here is that any concept of community, whether it is inshore coastal human communities or marine biotic communities (or their interrelationship), became anathema to the priorities of a rapidly industrializing fishery. Appropriate knowledge was conceived of in terms of development and production, and by their very definition, spelled disaster for both human communities and natural communities.

MAXIMUM SUSTAINABLE YIELD PRODUCTION MODELS

The fishery was in an expansion phase and the expansion was outstripping the science. Because there was no shortage of fish, there was no need for conservation. At least that is the way that the Canadian fishing industry saw it.

Edward Sanderman[13]

The assessment method that is used most often in commercially exploited fish is the relationship between catch and effort. In its simplest form, it is assumed that a given unit of fishing effort will result in a catch that is proportional to the fish abundance. The next and most crucial question for production model fishery scientists is: how many of the fish can be taken without impairing its future viability? It is this question that forms the basis for resource management in the fishery (as in any other industry based on a renewable resource), and for sustainability issues generally. What is different about the fishery is that it is very difficult to be absolutely sure you know how many fish there are in an industry at any one time. Based as they are on the catch and effort data of a commercial fleet, fish population in production models is an extrapolation based on the ratios of a controlled setting (provided by the stock assessment model) set against historical patterns of viability in the fishery. This is done through a standardized approach to effort that correlates the different methods of fishing into 'catch per unit of effort.' In other words, this many hooks longlining, equals this many miles of gillnet, equals this many hours towing a dragger net in this size boat. This was done by using the information in fishermen's logs concerning the number of days fished, where they fished, the gear used, the size of the boat, and the catches that resulted. Standardization was also used in the quantification of varying marine conditions and seasons into a 'catchability function.' Within many of the models there is also the assumption of standardized constants in relation to the ecological factors in the fish themselves.

One of the first concepts used in production models to set a limit on the size of the fishing industry was Maximum Sustainable Yield (MSY). First utilized in the 1950s by ICNAF in the management of the Northwest Atlantic, it was a modeling approach concerned only with inputs in terms of the biological production of fish and outputs in terms of fish caught. Models such as MSY are called production models because they focus on the idea of this annual surplus production of fish that can be exploited.

Often, historical records on catch and effort constitute the only information on the fishery, and production models were devised to relate yield to fishing effort. Because of the lack of information, these models were not concerned with actually assessing the size of the schools of fish, but rather with the more narrow concern of relating catch and effort.[14] Growth of the population of the fish was seen as a unified process of regeneration that was proportional to the size of schools, much in the same way that fishing effort expanded with the size of the fleet. These two factors were the input and output that had to be balanced if there was to be maximum sustainable yield. The number of fish produced over and above that needed for the replacement of the population was regarded as surplus and could therefore be exploited. To quote Rivard, "Total production has been defined here as an annual change of biomass resulting from somatic growth and recruitment in the exploitable portion of a fish population."[15] The goal of resource managers was to balance regeneration and exploitation. When this was achieved, the fishery was assumed to be in equilibrium.

Although somatic growth could be controlled to a great extent by restricting the fishing effort, the recruitment of juveniles to the migrating schools is at the mercy of environmental factors and largely beyond the control of managers. Even the most stable conditions in the marine ecosystem rarely lead to a constant recruitment for fish that are not exploited. It will vary in response to changes in temperature, current patterns and upwellings, as well as changes in the population of its predators and prey. In exploited schools of fish, recruitment is also affected by changes in the age structure of the fish and also by changes in fishing pressure. Despite the fact that it can vary greatly from year to year, recruitment is assumed to be constant over the long-term in production models, for the most part because fishery scientists didn't have the information to deal with it in any other way.

The 'catchability function' represents the changing marine conditions in a production model. It assumes that fish are equally distributed over the area in question, and attempts to even out seasonal differentiation. The production is therefore defined in terms of yield per unit of effort. Gulland (1961) attempted to increase the reliability of the production model by relating the yield per unit of effort on the fish in a given year, to the average fishing effort for previous years.[16] This was based on the assumption that the effort over a period of years represented an equilibrium effort. Like many attempts to modify the surplus production models so that allowance would be made for such things as biotic communities not in equilibrium, it was an attempt to include forms of randomness, like changes in age-structure, that would mitigate the use of so many constants in the models despite the lack of good data, and thereby render a more accurate assessment of fish population. Beddington and May (1977) attempted to modify the production model to include the effect of changing levels of fishing effort on biotic equilibrium. They did this by making the growth constant of the fish population subject to a random variable.[17] Schnute (1977) dealt with environmental randomness by analyzing how the surplus production model "fit" the fishing data and developed a "failure index" that urged caution in making resource management decisions in situations where the model did not successfully explain the data.[18] As a further attempt to refine the surplus production model, Walter (1978) incorporated a coefficient to reflect the annual changes in the recruitment of new individuals to the migrating schools.[19]

These alterations to the surplus production models were attempts by the scientific community to deal with the recognized complexity of marine ecology, and to incorporate that complexity into their models. This once again underlines the fact that classical production models assume that fishing is the only variable in an otherwise homeostatic environment. The variation in recruitment due to environmental changes and variations due to interactions with other species, represent obvious departures from this assumption.

Classical models also do not recognize the differences between age-groups. Instead, they treat fish populations as a whole. This does not take into account the effect of fishing efforts on a school, in spite of the fact that it has a profound effect on the age-structure of the fish. Also, the classical models make no provision for the migration patterns of the school, nor the ability of the fleet to track that migration. Commercial effort and fish are assumed to be equally spread over the grounds. This is contrasted with an increasingly efficient fleet that can track shrinking schools and maintain high catches in spite of a sharp drop in the fish population. Because effort and fish are assumed to be evenly distributed, the production model will give falsely optimistic predictions on the size of the schools of fish when exploited in the context of an increasingly efficient fleet.

In view of these limitations, production models were not considered to be more than a convenient, however crude, representation of a complicated process. They were, however, the only tools available for analyzing fish species where only catch and effort data were available. Rather than deal with the ecological complexity of the marine environment, production models functioned in a cycle of effort and yield set against past levels of fishing.

It is interesting to note that when one country (Canada) became responsible for preserving the fish and there were no longer competing interests on the high seas "common," the traditionally higher catches based on MSY production models were immediately deemed not to be a viable way of assessing catch limits. To quote from the DFO report, *Resource Prospects for Canada's Atlantic Fisheries 1980-1985*,

> There has been a major change in fisheries management approach within the past few years, particularly the abandonment of the Maximum Sustainable Yield (MSY) concept as the basis for establishing levels of harvest. The objective of MSY management was to obtain the maximum sustainable physical yield from the resource, i.e.., to get every available ounce of sustainable production from the fish stocks. This approach had serious drawbacks, not the least of which was the cost of getting that production.[20]

In other words, in the context of open access exploitation, depletion of the fish was an externality which everyone conveniently ignored as they maximized their own advantage. When the grounds became subject to a "unified directing power" after the 200-mile limit, the goal of management was to overcome this externality of depletion.

As noted in the above quotation, factors concerning the profitability of catching fish were added to the model of stock assessment so that the health of the fishing industry, as well as the health of the marine fish communities, was taken into consideration when setting the Total Allowable Catch. An alternative resource management concept called optimum sustainable yield (OSY) was developed which included social and economic, as well as biological considerations. Greater consideration was now given to the age-structure of the schools because of its importance as a factor in overall equilibrium. Also, as the commercial fleet modernized, it became much more difficult to standardize effort and therefore production models were less reliable predictors of fish population. Rather than derive the maximum yield from the fish, it was now the goal of fisheries management to maximize the difference between effort expended and benefit accrued. As the *Resource Prospects 1980-85* report states,

> For the moment, a somewhat arbitrary reference point which scientists call "Fo.1" is in wide use. In general terms, this corresponds to a level of fishing beyond which increases in total catch relative to increases in fishing effort are marginal.[21]

The fishing level of Fo.1 is set somewhere between MSY and maximum economic yield (MEY). To be more precise, the Fo.1 is the level of fishing mortality where the increase in yield due to adding one more unit of fishing effort is ten percent of the increase in yield due to adding one more unit of fishing effort in an unexploited school. It corresponds to a fishing mortality that "identifies approximately 20% of the total fishery for commercial exploitation each year."[22] The result is a value which is based on "catch per unit of fishing effort" (CPUE). Although the mathematical models became more subtle and complex, as well as being increasingly based on age-structure models, the concept of Fo.1 is still being used as a management indicator.

AGE-STRUCTURED ASSESSMENT MODELS

With the expansion of the monitoring programs in the late 1970s for most of the Northwest Atlantic groundfish, stock assessment relied more and more on age-structured models (also called analytical or dynamic models). This search for models which mimic more closely the ecological dynamics of a fish population, signaled a move away from production models. Contrasted to the production models which saw both the growth of the fish and the catch effort in terms of a unified rate, the science community was now beginning to use age-structure models that were based on the various processes that alter fish population, and had begun to deal with them as components in a more complex ecological environment. Within this framework, there is an attempt to incorporate the randomness of ecological factors which are not directly related to fishing mortality.

In the same way that the use of constants and ratios in production models no longer served to convey the realities of the marine environment, the use of a standardized value for fishing effort was made more problematic by rapid advances in electronic technology and new gear types, in the expanding Canadian fleet. Any such unaccounted for changes in the behaviour of the fleet towards the fish made effort comparisons unreliable. In a report on the 1979 assessment, D. F. Gray states:

> In 1978, about 97% of the catch went to the Canadian fleet. Much of this catch, particularly most of the portion caught by otter trawls was misreported as to catch location. Reported effort levels were also unreliable, and therefore there is no usable effort data for 1978.[23]

Even after he had revised the method for standardizing effort in the Canadian industry (which provided the basis for the models), fishery scientist S. Gavaris concluded:

> The model in its present form is best applied to the directed portion of a fishery. For some mixed fisheries the model would not apply. A solution to this important practical problem could be the basis for future research.[24]

Just about the only place where "catch per unit of effort" as an indicator of fish abundance is still being used, is in connection with the large off-shore draggers where fishing activity rarely varies and an accurate fishing log is kept. Nonetheless, CPUE is still considered a standard element in stock assessment, along with research surveys and age-structure modeling, and is referred to regularly in stock assessments up to the present. Despite having revised CPUE models several times since 1977, Gavaris — in the article in which he develops the ADAPT framework which is now used by DFO for stock assessment — stated that "There is no reliable information on effort from the commercial fishery."[25]

Within the Canadian context, fishery scientists began in the late 1970s to use age-cohort models such as the ones Beverton and Holt (1957) and Doubleday (1976, 1981) developed. Fish populations are composed of a number of discrete cohorts of age groups, and these can therefore be analyzed separately. Fishery scientists derived this information on the age-structure of fish populations by examining periodic samples of commercial catches. This bypassed the difficulty of obtaining a reliable standardization of effort, since the use of age-structure allows estimates of fish schools and fishing mortality to be made from catch data that is independent of effort measurement.

This sequential population analysis (SPA), as it is sometimes called, refers to the sum of a cohort of fish present in the water at any given time that are destined for subsequent capture in the fishery. This total is obtained by summing backwards from the terminal year (death), all the contributions that a particular year class has made to the fishery over its lifetime. This number is adjusted upwards to account for natural mortality. For example, all the fish that were recruited to the migrating schools in 1982 are considered to be the 1982 year class. Although it depends on the gear type, by about 1985, these fish will be large enough to be caught. The totals are then kept for three year olds caught in 1985, four year olds caught in 1986, and so on, until terminal age of fishing in about 1995. If this total is adjusted for natural mortality, what we get is a total that equals the number of fish recruited as exploitable fish in 1982.

This ends up being a little bit like predicting the past, but fishing levels have a great deal to do with precedents based on the effect of past fishing levels on the fish population. The information from SPA is also compared against current research vessel data and CPUE (if it is considered reliable), to determine if they are similar to other models in terms of rendering a comparable view of the state of biotic communities. Because Beverton and Holt

(1957) showed that the age of recruitment to the population and the age of the availability of the cohort to the fishing gear are independent variables, resource managers were able to find the optimum age a fish should enter the fishery in combination with a certain level of fishing that would allow the largest yield with the least impact on the growth of the fish population. For example, if the mesh size of dragger nets is increased, the age at which a fish enters the fishery also increases. This leaves the fish unperturbed for a longer period, which increases both the size of the fish population and the number of eggs produced, thereby affecting recruitment numbers. This change in gear type could possibly allow increases in the level of fishing if the biotic community was in a viable state. It also follows that under steady-state conditions (i.e. stable age distribution), the annual catch from the overall population under a given fishing strategy equals the catch that can be taken from a single cohort throughout its life under the same strategy.[26] This has been called the "principle of equivalence."

There is a simple equation that describes the four main processes taking place in an exploited school of fish:

$$S(1) = S(2) + (R+G) - (M-F)$$

$S(1)$ and $S(2)$ represent a time interval, say one year, R is the recruitment of new individuals, and G is tissue growth of the existing school of fish. Subtracted from this is mortality from fishing (M) and reductions in the size of the population (F), from events such as predation, natural mortality, and environmental factors (ranging from el Nino to oil spills). Even the most complex dynamic or age-structure model can be summarized in terms of this equation. It only becomes complicated when sub-models are created to realistically reflect the four processes. These models are achieved through a compromise between realism and precision so that the model can stand as a reasonable simulation of the processes going on in the system while responding to the available data.

This perspective on how these four processes affect different age-classes is reflected in *Resource Prospects for Canada's Atlantic Fisheries 1989-93* published by DFO:

> In practice, fish stocks vary in abundance from year to year, both in absolute numbers and in the availability to fisher-men. Their abundance depends upon the balance between the number (or weight) of young that enter the population during the year, the growth of the individuals in the population, the losses due to natural causes (predation, etc.), and the removal due to fishing. None of the natural events are constant from year to year and the mortality of the very young is particularly variable. Hence, there is considerable fluctuation in the number of small fish reaching harvestable size in any year, while the rates of growth and mortality at all ages will also vary. Using recent biological information to predict how much fish will actually be caught is complicated further by variation in the availability of the fish to fishermen due to annual differences in migration routes, areas of distribution and degree of concentration.[27]

This explication of the difficulties of stock assessment, although it verges on crying the blues, is an accurate description of the way DFO viewed the marine environment, and the difficulties it faced in analyzing the size of fish populations.

The great advantage of the age-structure model over production models which use a standardized catch rate, is that it can predict the effect of different fishing strategies on the fish population, as well as on each individual age-class. Also, because the model deals with each age-class individually, it can be employed to simulate the exploitation history of the fishery, as well as predict its age-structure in the future. For resource managers to produce an annual fishing strategy, the fishing mortalities-at-age are applied against the present school abundance-at-age derived from cohort analysis. The calculated removals-at-age multiplied by the mean weight-at-age provides the Total Allowable Catch. If this is calculated at the Fo.1 level, it will correspond to roughly 20% of the fish population. These catch projections are normally the last step in the stock assessment process. They represent the core of the information provided to the fisheries managers and politicians by the assessment biologists, and form the basis for setting the Total Allowable Catch.

RESEARCH VESSEL SURVEYS

In terms of research vessel activity, the controlled setting which provides the basis for the assessment of fish populations is linked to the towing patterns of the net. In the catch and effort data of the commercial fishery, it is the stock assessment models that create a controlled setting from which extrapolations can be made concerning fish populations. The information that is derived from the mathematical models based on commercial data, provides biologists with insights into the exploited sectors of the marine ecology. Beginning in 1970, the Canadian government has also utilized standardized bottom trawl surveys (usually once a year) to analyze the condition of the fish. Research surveys have several advantages over commercial data: (a) they do not have a bias arising from non-random distributions of fishing locations (i.e. they are not chasing the fish the way commercial fleets are); (b) there is a more accurate division of the fish in terms of age-classes; (c) by using different mesh size in their nets, they can assess age-classes outside of the exploitable range, as well as aspects of the food chain such as ichthyoplankton.[28] Whereas recruitment of juvenile fish to the migrating school is assumed to be constant in models based on commercial data, it is possible for research vessels to assess the survival of fry larvae. Recruitment is the single most important factor outside of fishing mortality for determining fish populations. However, it remains outside the control of managers, and until the use of research vessels, outside of their mathematical projections.

Canadian survey vessels utilize the stratified random design for towing. Within each stratum (or depth), the sampling unit (net) was defined by an area swept by a trawl with a door of a specified width, for a distance of 1.75 miles. The trawl's door multiplied by the distance covered, will give a value that describes the area as a whole, which can then be compared against the quantity of fish caught. The actual sampling frame for each strata consists of 5' latitude by 10' longitude rectangles, each subdivided into a number of trawling locations. The choice of sample units is done by first randomly selecting a rectangle and then randomly selecting a location within that rectangle.[29] The surveys are also done at the same time every year in an attempt to standardize the migration of the schools. Once the estimates for the total number or weight and stratified mean catch have been obtained for a particular trawl survey, the next step is to relate these estimates to the actual population size. The assumptions required to make this correlation are as follows: (a) All the sizes of the fish of a particular species are equally catchable. This implies that the size composition in the survey catch is an unbiased estimator of the size composition of the target population as a whole. (b) No fish escape from the net while it is being towed. (c) The net behaves the same under all conditions (bottom types and sea and tide conditions).

The results of these surveys are used as a series of estimates carried out over a number of years. Over time, trends found in these estimates are expected to reflect trends in the population. These trawl survey values are then compared against the data from commercial catch rates from the same schools. Because of the difference in volume, non-randomness, and gear type (between research and commercial models), there were no direct proportional comparisons until the ADAPT framework was developed. Any relationship could only be inferred from long-term averages.

Despite the advantages of the research vessel surveys in predicting fish populations, there are also questions about them in scientific terms. As Finlayson states:

> One might assume that data from this source [research vessels] would be the most rigorous and least ambiguous. The data is collected directly by scientists through research designed expressly for that purpose. In fact, scientific popula-

tion research surveys do not appear to be any less susceptible to constructed inputs than other data sources. . . . The first is that, by normal standards of statistical validity, the population is hugely undersampled. . . . Research vessel operations are very expensive, time consuming, and must be negotiated in competition with other demands for available financial and human resources. . . . And the validity of a projected population portrait derived from such a small slice of space/time is the subject of considerable debate.[30]

Once again, it is clear that since fish only appear in the context of human activity, finding a way to assess the number of fish in the ocean based on human activity, even when it is not economic activity, remains very difficult.

THE ADAPT FRAMEWORK

An adaptive framework model for the estimation of population size currently used by DFO was developed by Gavaris. As he describes it, the model is

. . . based on minimizing the discrepancy between observations of variables and the values of those variables predicted as functions of population parameters, and provides a statistical basis for this type of problem. The flexibility in the types of data and relationships which may be employed is considered essential in order to handle the wide range of situations encountered in stock assessments.[31]

Its main function is concerned with "tuning" the results of commercial catch-at-age data, research vessel data, and CPUE statistics, if they are applicable. What it did was compare the commercially based age-specific information directly with that of the research vessel, rather than comparing them indirectly through formulae, as had been done before. This method minimizes the discrepancies between the two sets of data and allows the assumptions underlying the method to be more carefully questioned.[32] This same adaptive framework is also used to "tune" SPA to CPUE, when applicable. From a statistical standpoint, ADAPT provides a better fit to the data and is presumed to be more reliable than previous models. The perspective behind the development of these models was described by Gavaris as follows:

> The development and implementation of an adaptive framework was based on the observation that an assessment technique is composed of two parts, a method of estimation and the definition of a model. For the method of estimation, a statistically established method for solving the problem of estimating the parameters of a model was selected. With respect to the definition of a model, it was recognized that the variety of situations encountered would require considerable flexibility. The philosophy of an adaptive approach was taken rather than attempting to identify a model which would perform "well" in "typical" situations.[33]

This attention to the development of new models and the acceptance, almost in passing, of a "statistically established method" was commented upon by Harris:

> Certainly, the ADAPT model is an improvement upon the model used in previous years, but we are somewhat concerned that the analytical process is being overemphasized while insufficient attention is given to the quality of the data inputs.[34]

As a component of economic development, "the quality of data inputs," returns discussion to the social and economic context in which "data" exists. In the case of the Northwest Atlantic fishery, the "adaptation" for fishery scientists was in becoming a service industry which attempted to understand the inputs and outputs of a biotic community in a state of constant flux, while at the same time satisfying economic demands from a commercial industry desperate to maintain its share of a disappearing raw material.

CONCLUSION

> The decision was that the cost of a liberal error (i.e. too high a quota) was that we slowed down the rate of growth (just what did happen in the mid-1980s), but did no damage to the stock. The cost of a conservative error was that we shut down thousands of jobs and caused the associated hardships for nothing. The choice made is history.
>
> Jake Rice[35]

Despite attempts by the ADAPT model to "tune" the various snapshots of the health of fish communities, there was a profound discordance between models based on commercial catch data and those from research vessels. While the research vessel analysis gave every indication that the fish were in serious ecological trouble, the models based on commercial catch data showed the fish to be in a relatively stable state. It was this discordance between the models which led to the Harris Report, and eventually, to the indefinite fishing moratorium being declared in July, 1992. As the Harris Report states:

> . . . the data themselves are still to some considerable degree unreliable or, at least, subject to strong suspicion of unreliability; and, this stricture applies, though perhaps not with equal force, to both the Research Vessel data and the catch per unit of effort data. The former, it is believed, might be improved through increased sampling effort, by appropriate correction for time of survey, and for environmental variabili-

ty. The latter are, perhaps, distorted by underestimation of
the significance of technological changes in catching effectiveness when fishing is conducted primarily upon spawning
or other aggregations.[36]

The navigational range of the fleet, as well as its electronic capability
to track fish — many times waiting until they were schooled up to
spawn before exploiting them — appeared in catch and effort data as
fishing for a relatively short time and catching a reasonable amount of
fish. In the aftermath of collapse, it became clear that the ability of
the fleet to chase fewer and fewer fish, all the time maintaining prof-
itable catches, never sufficiently entered the modeling process, and
gave mistakenly optimistic view of the health of biotic communities.
Even after the moratorium was declared, the large companies — who
were the ones who supplied the catch and effort data — still claimed
there were lots of fish to catch. In other words, and in the final analysis, conservation, as defined here, is an economic question, not an
ecological one. This economic question has more to do with technological capability affording an increasingly intrusive entry into the
natural world, which renders any conception of nature outside of that
context — even the one provided by inshore fishers — as external to
the equation.

DFO scientist D. Gascon conveys the grim state of fishery science as it attempts to operate in the intensifying vortex of collapse:

> Although it is conceivable that future improvements in
> catch projections may take place on the form of more
> sophisticated models including growth functions,
> recruitment functions, etc., the present degree of
> imprecision in the data seems to preclude this kind of
> development in the immediate future. Rather, current
> trends seem to indicate a move toward simpler models
> with a greater degree of reliability.[37]

The conditions in the industry as a whole, and most significantly, the antipathy between the fishing fleet and DFO which resulted in gross misreporting, affected the science of the biologists by limiting the availability of reliable data. The paranoia and conflict endemic to the situation is graphically illustrated in a DFO report on the status of the cod on the Scotian Shelf by Campana and Hamel:

> A more serious misreporting problem first became apparent in 1986 and continued into 1988; comments by fishermen, industry representatives, and port samplers indicate that substantial quantities of cod were either unreported during the year or incorrectly reported as other species such as white hake. It was impossible to quantify the magnitude of this underreporting, but its extent appeared to greatly exceed that of previous years. Therefore, reported catches may underestimate actual catches by anywhere between 10-40%. . . . Given the misreporting problem in the region, the multiple closures and fishery restrictions in recent years, the low proportion (7%) of catch entered into the model, and the low number of observations per model category, the model results were not considered to be useful. In addition, the use of commercial CPUE as an index of 4X [a fishing zone] cod abundance was deemed inappropriate.[38]

This assessment indicates the extent to which the relationship between DFO and the fleet sector has disintegrated. Declining catches and the imposition of more stringent quotas, coupled with fleet overcapacity driven by the expansionist dynamic of market economy, led to an antagonistic situation where there was wanton and consistent misreporting by boats in order to avoid quota restrictions. In fact, the single most important realization that comes out of the Harris Report is that the rapid expansion of the capacity and efficiency of the Canadian fleet did not enter the scientific frame of reference and, as a result, skewed the models. By the time the moratorium was declared in 1992, it became clear that increased catches were not due to the growth of fish populations, but rather in the growth of catching technology, which continued to catch substantial amounts of fish even as the populations were collapsing.

Within this social and economic milieu of conflict and mistrust caused by ecologic collapse, the knowledge claims of fishery science have been substantially reduced. As Finlayson states of the re-evaluation which occurred in the aftermath of collapse:

> Stripped to bare essentials, all that science is now willing to claim is that too much Northern Cod is being caught and that quotas should be reduced to protect the resource, exactly what the inshore fishermen have been saying for some years.[39]

As it turns out, only now when they are all gone are the fishery scientists sure how many fish are in the ocean.

Wolfgang Sachs conveys the transformation in the perspective on the natural world within environmentalism generally, as a shift from ecology — which "carries the promise of reuniting what has been fragmented, of healing what has been torn apart, in short of caring for the whole" — to ecosystem theory, which is based on "the science of engineering feedback mechanisms" and on the metaphor of a "self-governing machine."[40] A project that began in the name of conservation, ends up as the facilitator of more intense forms of economic development. As Sachs states:

> Today, however, the responsiveness of nature has been strained to the uttermost under the pressures of modern man. Looking at nature in terms of self-regulating systems, therefore, implies either the intention to gauge nature's overload capacity or the aim of adjusting her feedback mechanisms through human intervention. Both strategies amount

to completing Bacon's vision of dominating nature, albeit with the added pretension of manipulating her revenge.[41]

This transformation in views of nature is precisely what happened to fishery science in the Northwest Atlantic. What was to have been the "focus on conservation" became the conceptual counterpart to economic development. By attempting to represent natural communities in the profoundly impoverished terms of production and exploitation, fishery science entrenched and confirmed a view of natural communities which insured their destruction within the hegemony of modern economy. The "focus on conservation" turned out to be a service industry for industrial development.

NOTES

1   Bruce Mitchell. 1989. *Geography and Resource Analysis*. London: Longman & Wiley, p. 3.

2   Jake Rice quoted in Alan Christopher Finlayson. 1994. *Fishing for Truth: A Sociological Analysis of Northern Cod Stock Assessment from 1977 to 1990*. St. John's: Institute of Social and Economic Research, p. 149.

3   Leslie Harris. 1989. *Independent Review of the State of the Northern Cod Stock*. [Interim Report. 1989, Final Report. 1990] Ottawa: Minister of Supply and Services, p. ii.

4   Harris. (1989:ii).

5   Fisheries and Marine Service. 1976. *Policy For Canada's Commercial Fisheries*. Ottawa: Department of the Environment, p. 63.

6   Finlayson. (1994:6).

7   Finlayson. (1994:24).

8   Finlayson. (1994:25).

9   Leslie Harris quoted in Finlayson. (1994:69).

10   Edward Sandeman quoted in Finlayson. (1994:85-86).

11   J. J. Maguire quoted in Finlayson (1994:97).

12   Edward Sandeman quoted in Finlayson (1994:110).

13   Edward Sandeman quoted in Finlayson. (1994:86).

14   Rivard, D. 1988. "Biological Production and Production Models." Collected Papers on Stock Assessment Methods. [D. Rivard ed.] CAFSAC Res. Doc. 88/61 44-110.

15   Rivard. (1988:48).

16   J. A. Gulland. 1961. "Fishing and the stocks of fish at Iceland." *Fishery Invest. Lond. Ser*. II, 23: 1-52.

17   J. Beddington & R. M. May. 1977. "Harvesting Populations in a random ly fluctuating environment." *Science 197*: 463-465.

18   J. Schnute. 1977. "Improved estimates from the Schaefer prod uction model: theoretical considerations." *J. Fish. Res. Board Can*. 34: 583-603.

19   G. G. Walter. 1978. "A surplus yield model incorporating recruitment and applied to a stock of Atlantic mackerel." *J. Fish. Res. Board Can. 35*: 229-34.

20   Department of Fisheries and Oceans. 1979. *Resource Prospects for Canada's Atlantic Fisheries 1980-1985*. Ottawa: Department of Fisheries and Oceans, p. 2.

21   DFO. (1979:2).

22   Harris. (1989:1).

23   D. F. Gray. 1979. "4VsW: Background to the 1979 assessment." *CAFSAC Res. Doc.* 79/30, p. 3.

24   S. Gavaris. 1980. "Use of a multiplicative model to estimate catch rate and effort from commercial data." *Canadian Journal of Fisheries and Aquatic Sciences*. Vol. 37, No. 12, p. 2274.

25   S. Gavaris. 1988a. "An adaptive framework for the estimation of popu lation size." CAFSAC Res. Doc. 88/29, p. 4.

26   M. Sinclair, R. O'Boyle, & T. D. Iles. 1981. "Consideration of the stable age distribution assumption in 'analytical' yield models." *CAFSAC Res. Doc.* 81/11.

27   Department of Fisheries and Oceans. 1988. *Resource Prospects for Canada's Atlantic Fisheries 1989-1993*. Ottawa: Department of Fisheries and Oceans, p. 3.

28   S. Gavaris. 1988b. "Abundance Indices from Commercial Fishing." *Collected Papers on Stock Assessment Models*. [D. Rivard ed.] CAFSAC Res. Doc. 88/61, p. 4.

29   S. J. Smith. 1988. "Abundance Indices from Research Survey Data." Rivard. (1988:16-43).

30   Finlayson. (1994:72).

31   Gavaris. (1988a:2).

32   Harris. (1989:17-18).

33   Gavaris. (1988a:8).

34   Harris. (1989:35).

35   Jake Rice quoted in Finlayson. (1994:83).

36   Leslie Harris. 1990. *Independent Review of the State of the Northern Cod S tock*. Ottawa: Minister of Supply and Services, p. 3.

37   D. Gascon. 1988. "Catch Projections." Rivard. (1988:135-146).

38   S. Campana & J. Hamel. 1989. "Status of the 4X cod stock in 1988." *CAFSAC Res. Doc.* 89/30, p. 3-4.

39   Finlayson. (1994:147).

40   Wolfgang Sachs. 1992. "Environment." *The Development Dictionary*. London: Zed Books, p. 32.

41   Sachs. (1992:32).

V

Common Property and the Enclosure of the High Seas

Economics and Ecology. . . are intellectually indistinguishable as far as dealings with nature are concerned — but only if all costs are fully taken into account. The free enterprise system is devoted to doing just that — making sure that all costs are accounted for, so that no border crossings, trespasses, or violations of private property rights takes place.

W. E. Block[1]

The competitive individualism that Hardin believed inevitably led to "the tragedy of the commons" might not be a generic failing of the human species but rather the specific historical consequence of the social changes that followed the advent of modern capitalist modes of production and social organization.

Arthur McEvoy[2]

A discussion of common property provides an evocative entry point for a discussion of the regulation of the high seas generally, and the Northwest Atlantic fishery in particular. In the first case, this is because common property is continually identified in ocean management literature as the main source of the many problems related to overcapacity and inefficiency, and the perceived failure of management goals to mitigate these conditions. Secondly, common property is repeatedly mistaken for a situation which would be better described as open-access capitalism, and points up the way most current discussion of environmental problems universalize modern economic relations. This makes it difficult to problematize the modern structures of everyday life when discussing these issues.

I have described the evolution of management frameworks in the Northwest Atlantic fishery as moving from common property to public property to private property. Within this transformation, common property is conveyed in terms of where the world has been, and privatization is where the world is going. In this sense, it is possible to describe it as an enclosure movement which — although claims to the high seas were continually defined in terms of promoting conservation — in reality facilitated the intensification of industrial exploitation, much in the same way as the enclosure movement on land had done earlier in the modern period in Europe. This enclosure movement in the fishery also promoted the same displacement and marginalization of certain groups who were not essential to the industrialization of relations. So, I believe it is important to see the connection between the enclosure of the oceans and the intensification of industrialization related to the expanding forces of globalization. Although coastal zones were nationalized, it is clear that national mandates are increasingly serving these global realities, rather than resisting them.

In response to the expanded human enterprise that came with industrialization generally, and the industrialization of the fishery in particular, the natural world has been transformed into resources which had been held in common, to those subject to the more permanent dominion of private ownership. This transformation in relations associated with the expansion of modern economic and technological realities, highlights the significance of corresponding institutional and legal frameworks such as property rights, which support those realities. In other words, the forces that lead to the destruction of the natural world as reflected in a range of modern social and economic institutions. Within these institutions, there is no space for challenging the unified voice of exploitation and thereby conserving the natural world.

A very interesting aspect of the debate over common property relates to the recognition of the historical moment in which this debate appears. It can be argued that cultures who practice common property in social anthro-pological terms — as the historically-specific social and cultural institutions which control access to, and the use of, a shared benefit, as well as provide the rationale which informs the demands made on that benefit — have no recognition that they themselves are putting into practice, common property resource management. Common property can be seen as an oxymoron. The North American aboriginal responses to the various attempts to negotiate treaty lands provide some understanding of how puny and insulting the con-cept of property is to those who inhabit a "common" world. In fact, this debate over common property requires the existence of private property, as well as the appearance of overexploitation problems for there to be an accom-panying discussion of the common property "problem." As Durrenberger and Palsson state:

> For common property to be defined, it must be problematic.
> . . . This is likely to happen only under the pressure of
> overuse of the resource, which is the beginning of the tragedy
> of the commons. Such overuse is a function of trying to pro-
> duce more from the resource than it can yield over the long
> term. The conditions for such overproduction are found
> when production is organized for exchange rather than for
> use, a phenomenon of stratified societies.[3]

This statement points to the importance of the broad social context in which particular issues become discussable. When we assess the problematic nature of common property, this usually means that it is problematic for those who want to intensify both ownership and industrial production, and who sense the impediment posed by groups in a society who share a commons, as with the Scottish crofters. Common property is not problematic for crofters, or for artisanal fishers around the world. It is, or was, the social basis for their lives. But although it is not problematic, it is certainly under threat. So it is very important to be clear here that those who portray common property as a problem, are those who universalize modern forms of exchange related to capital. This is most obvious in Garrett Hardin's "tragedy of the commons" metaphor, where "every particular decision-making herdsman" operates in a historical vacuum which universalizes the level playing field of capital.

As outlined in the introduction, using the fishery as a case in point to assess the relationship between conservation and development with regard to sustainability, raises important issues concerning the destruction of the natural world. This relation between conservation and development becomes even more evocative with regard to the fishery when it is focused on the idea of common property. The answer to the question "Why is there strife, conflict, and no fish?" has tended to revolve in a very close circle around the problems associated with the common property nature of the fishery, and the oceans generally. It can therefore provide an interesting avenue into discussions about the relation between conservation and development.

Ridding the fishery of common property has been one of the major goals of Canadian government policy since it nationalized the Northwest Atlantic in 1977. In the name of increased efficiency and the rationalization of operations, resource managers have attempted to control what they perceived as the expansionist logic of common property, through a program of catch quotas, area closures, gear and vessel restrictions, limited-entry licensing, and the granting of property rights. A series of quotations will illustrate the chorus of complaints that have been leveled against the common property "problem" in the Northwest Atlantic fishery:

Fisheries and Marine Service, *Policy for Canada's Commercial Fisheries* (1976):

> In an open access, free-for-all fishery, competing fishermen try to catch all the fish available to them, regardless of the consequences. Unless they are checked, the usual consequence is a collapse of the fishery: that is, resource extinction in the commercial sense, repeating in a fishery context "the tragedy of the commons."[4]

Parzival Copes, *Canada's Atlantic Coast Fisheries: Policy Development and the Impact of Extended Jurisdiction* (1977):

> Economists have traced the special problems of the fishery to the circumstance that it depends on utilization of a common property resource. . .Where no property rights are exercised with respect to the resource, there is a tendency for excessive manpower and capital to enter the industry. This leads to dissipation of the rents the industry could earn and to low returns for labour and capital.[5]

Kirby Task Force, *Navigating Troubled Waters: Report for the Task Force on the Atlantic Fisheries* (1983):

> It must be further acknowledged that the fishery is rife with incentives to expand. The most important of these are the common property nature of the resource.[6]

E. Weeks and L. Mazany, *The Future of the Atlantic Fisheries* (1983):

> The common property aspect of the fisheries is fundamental to many of the current problems of the fishery.[7]

Haché Task Force, *Scotia-Fundy Task Force Report* (1989):

> The familiar "race to fish" leads to too much investment in fishing capacity, which puts the fishing industry in a fragile financial state even when fishing is controlled well enough to protect the stocks.[8]

Cashin Task Force, *Charting A New Course: Toward The Fishery Of The Future* (1993):

Too much is being demanded of the fishery. . . . A basic reason is that a fish swimming free in the ocean belongs to anyone with a legal right to catch it. It is the common property of everyone holding an applicable license. Everyone wants as much of the resource as possible. So, without adequate controls, the resource declines. And, if more people can put pressure on the resource, average returns decline even faster.[9]

L. S. Parsons, *Management Of Marine Fisheries in Canada* (1993):

Within Canada over the past 25 years a complex web of measures has been developed to deal with the problem of the "race for the fish" [related to common property]. These include allocation of access. . . limited entry licensing. . . and individual quotas. Individual quotas are a step in the evolution of property rights.[10]

Angel et al., *Report of the Workshop on Scotia-Fundy Groundfish Management from 1977 to 1993* (1994):

Groundfish resources in Canada have traditionally been treated as common property. . . where none own a particular share of fish. This basic characteristic creates what is called the common property problem . . . [which leads to] overexploitation. . . , is economically wasteful. . . , [and causes] competition and social conflict.[11]

This was the expansionist dynamic that was identified as the source of the problems of an unregulated fishery, and it was this dynamic that the Canadian government hoped to bring under control with a comprehensive management framework when it extended its coastal economic zone to 200 miles in 1977. It was the goal of the "unified directing power" identified by Gordon, to mitigate the inefficiency and overcapacity in the fishery.

Although it was greeted with great enthusiasm following the declaration of the 200-mile limit, the public property challenge to common property based on comprehensive regulation of a "unified directing power," was perceived to have failed throughout the 1980s in the Northwest Atlantic. As a result, the Canadian government gradually began to grant ownership of a share of the annual quota to the larger participants in the industry, in hopes that privatization would stem the depletion of the fish and promote economic stability, while ridding the fishery of ongoing conflict between exploiters and regulators which had been generated by public property management approaches.

It is the contention of this book that the problem of over-exploitation goes beyond one of property rights, and as with the discussion surrounding sustainability, what needs to be made problematic is not common property, but the structures and processes of modern life which support the expansion of capital and make a dead thing grow. Solutions that move the fishery in the direction of privatization need to be reconsidered in light of the widespread environmental problems under various forms of property rights throughout the world. A clarification of the problems of overcapacity and depletion which address the growth of a dead thing, would alter the discussion and move it toward ridding the fishery of the pressure of capital theory, rather than of common property. If it is accepted that there is a connection between the depletion of the fish in the Northwest Atlantic and a range of environmental problems throughout the world (in terms of what causes these problems), then it becomes apparent that there are a larger set of interconnected issues that have more to do with modern economic and technological realities, than they do with common property.

Capital abhors all communities but itself, whether it is common land in Southern Mexico, social welfare programs, or cod fish. The open access which decimated the fish in the world's oceans is not unlike the privatization, deregulation, and free trade platforms of economic globalization. The "common property problem" is therefore the expression by those who have an abhorrence for any community but the community of capital.

RATIONAL TRAGEDY

If the goal of this chapter is to look at common property in the fishery so as to better understand the relation between conservation and development, it is important to clarify some of the confusion surrounding the use of the term. Common property has had its most widespread use in relation to *The Tragedy of the Commons* by Garrett Hardin.[12] In his essay, Hardin attempts to set out the perils associated with the free and unregulated access to scarce resources by focusing on the commons and the expansion of human population. If this exponential increase in human numbers takes place "without relinquishing any of the privileges we now enjoy,"[13] there will be an inexorable depletion of the Earth's resources. The recurrent theme in Hardin's work is the remorselessness of this mathematical equation:

> Each man is locked into a system that compels him to increase his herd without limit - in a world that is limited.[14]

> . . . we are locked into a system of "fouling our own nests" so long as we behave only as independent, rational, free-enterprisers.[15]

> To couple the concept of the freedom to breed with the belief that everyone born has an equal right to the commons is to lock the world into a tragic course of action.[16]

> Individuals locked into the logic of the commons are free only to bring on universal ruin.[17]

In Hardin's terms, there is no technical solution to the relation between individual freedom and environmental problems, so long as there is not the acceptance of "the infringement of individual liberty." For Hardin, this infringement is represented first by the extension of property rights "enclosing farmland and restricting pastures," then the restriction of pollution through "coercive laws and taxes," and presently, the "recognition of the evils of the commons in the matter of pleasure" with regard to the right to procreate freely. The expansion of the human population poses, for Hardin, the most serious threat to survival. What Hardin seems not to take into account is that this "locked-in" inevitability does not occur in a vacuum, as mentioned earlier, but requires specific social and economic realities for it to become problematic.

It is clear how the metaphor of the commons outlined by Hardin can be readily applied to the problems in the fishery. The overcapacity, inefficiency, and overexploitation that have plagued the fishery can be explained within Hardin's metaphor. Fishery managers have also taken Hardin's recommendations for overcoming the problems of the commons. First, by putting in place the coercive laws and property rights that would circumscribe what has been repeatedly referred to as "the rush to fish" caused by common property resource exploitation, and then, by privatizing shares of the commons.

In both fisheries discussions and in Hardin's presentation of the "tragedy of the commons," there is a tacit acceptance of the workings of market economy. It therefore becomes the universalized paradigm in which problems of the fishery and the commons are analyzed. By emphasizing individualism and competition, Hardin's presentation of the tragedy of the commons accepts the workings of modern economy as a universal form of human behaviour. This leads Hardin to single out what does not fit into the market economy paradigm — which is the commons — and to blame it for the overcapacity of the human enterprise. More recently, economists such as R. J. Smith have echoed the same view:

It is not private ownership that has led to the near extinction
of many species of whales but rather the very concept that
environmentalists cherish: common property "ownership" —
the idea that whales belong to everyone.[18]

What Hardin is really articulating in the "tragedy of the commons" is
the operation of an open access unregulated laissez-faire form of market
economy. The important contribution he makes to understanding this
growth-oriented, expansionist dynamic, is the recognition that there is no
outer limit of exploitation and depletion to the human enterprise in this context. Because he believes this expansion is caused by an open commons,
Hardin sees the solution to these problems in terms of the implementation of
legal restrictions and property rights. Although he recognizes the need to
limit human activity, Hardin articulates the problem in such a way that his
solutions remain within the market economy paradigm. This repeated confusion between common property and open access can be linked with the hegemonic power of market economy. As John Ferguson states:

The confusion between common property and open access
situations continues in spite of a long list of authors who have
attempted to distinguish them. . . . While this is a distinction
which has been articulated by many for well over two
decades, it remains on a kind of auto-pilot, especially in the
neo-classical economic literature.[19]

If the fishery is a case in point with regard to the "commons,"
property rights and restrictive measures will not necessarily overcome
the overexploitation of the natural world. As I have stated with regard to
the sustainability debate, the call to integrate conservation and development does not make development sufficiently problematic. In the common property debate, there is a similar reluctance to make development
problematic. The central problem relates to the "openness" of development, and not to the commons. It is the "openness" of modern market

economy which both national enclosure and privatization support, rather than restrict.

These kinds of reservations about Hardin's analysis have led to a re-evaluation of the scenario of the "tragedy of the commons" within such disciplines as anthropology and development studies. To quote Bromley and Cernea, who engage in an analysis of the past failures of World Bank projects:

> . . . common property regimes have attracted consider-able analysis and debate, with both researchers and development practitioners distancing themselves more and more from the stereotype of the "tragedy of the com-mons.". . . Common property carries the false and mis-placed burden of "inevitable" resource degradation that instead has to be causally attributed to situations of open access. . . . By confusing an open access regime (a free-for-all) with a common property regime (in which group size and behavioral rules are specified) the metaphor denies the very possibility for resource users to act togeth-er and institute checks and balances, rules and sanctions, for their own interaction within a given environment.
>
> The Hardin metaphor is not only socially and culturally simplistic, it is historically false. In practice, it deflects analytical attention away from the actual socio-organizational arrangements able to overcome resource degradation and make common property regimes viable.[20]

Vandana Shiva is especially acute in her analysis of the colonial approach of the World Bank to the institutional arrangements of developing countries. Also, she sees Hardin's metaphor as very much a part of an approach which universalizes modern rational, patriarchal, perspectives:

There is, of course, the popular triage thesis that the poor have no right to survival and should be dispensed with. Hardin's tragedy of the commons scenario emerges from male reductionist assumptions about nature and the logic of triage that such reductionism and its principles of exclusion and dispensability entail. Hardin is just a symbol of the new trend in reductionist science which uses the language of ecology and conservation to unleash another attack of violence against nature. More centralization, more uniformity, more manipulation become new and false prescriptions for overcoming the ecological crisis. Yet neither nature nor people can be saved when the destruction of the former and the dispensability of the latter are the presupposition for creating the new reductionist science of nature.[21]

Koula Mellos sees similar problems with rationality and science in the institutional arrangements and "coercive laws" which limit exploitation in the modern context:

> Social, political, and economic policy take their directives from science. . . . Scientific knowledge is the recognition of natural necessity and as such is the sole source of good and hence lays claim as the sole guiding force of moral social activity.[22]

As stated earlier, it is questionable whether a centralized agency relying on scientific information can promote the living diversity of the Earth. In the case of the fishery, a centralized government agency administering a state-owned fishing grounds has not proved workable. The complex web of regulations put in place to restrict access to the fish have not been able to control overcapacity, inefficiency, and depletion. McEvoy points to a possible reason for this failure:

> Progressive conservation envisioned a powerful central state made up of impartial experts who would command a passive citizenry to obey efficient laws. Rather than correcting the market failures that might lead to Hardin's tragedy of the commons, however, the structure and processes of lawmaking for the fishery merely duplicated them in a different forum.[23]

The fact that in 1977, resource managers justified expelling foreigners in the name of conservation, and then turned around and repeated the same process in the national context which had collapsed fish communities in the international arena, supports McEvoy's contention that the approach to conservation based on a central state and efficient laws (in terms of a market economy), is unable to limit the exploitation of the natural world. Rather than recognizing that it was the perpetuation of market economy realities within the regulatory arrangements which caused the collapse of marine communities in Canada during the 1980s, fishery managers came to believe that it was the lack of individual property rights among the participants of the fishery that had made the fishery difficult to regulate and which led to the depletion of fish species. This same privatizing response to conservation failures is taking place throughout the developing world. To quote Shiva:

> Yet the Wasteland Board schemes will primarily privatize the commons by transferring rights and control from the community as a whole, to the World Bank, private business, and a few local people.[24]

Once again we see an attempt to align social and economic realities to the market economy paradigm, which we have identified as the source of overexploitation. In the process, the people who had previously used the common "wasteland," but were not necessary for the industrialization process, suffered the consequences of dislocation and marginalization.

Other works such as *The Question of the Commons*, edited by Bonnie McCay and James Acheson, also focus on the re-evaluation of Hardin's metaphor:

> Scholars in many disciplines, ranging from economics and psychology to biology, have explored the dilemma of the commons and debated its solutions. Anthropologists, too — in their studies of subsistence economics, cultural ecology, property rights, law and social evolution — have addressed the issue of common property, bringing with them a rich tradition of inquiry into the relations between human groups and natural resources.[25]

This re-evaluation is set in the context of the problems plaguing the late twentieth century:

> The second half of the twentieth century may someday be recalled as the time that we became painfully aware of the social and ecological costs of industrialization, rising populations, and unsound resource management. . . . Problems such as those outlined above are explicable as tragic outcomes of common property tenure. . . . According to the theory popularized by Hardin, all resources owned in common- air, oceans, fish, grasslands and so on- are or eventually will be overexploited. . . . Hardin's most important message was that we cannot rely on normal market forces nor on people's best intentions to save their environments and themselves. . . . In the 1960s and 1970s. . . the only thinkable solution to commons dilemmas was government intervention. In the 1980s. . . the same problems and the same theory trigger discussion of another solution: privatization.[26]

In either case, it is assumed that the users themselves will not change the system, and that a centralized authority must do it for them. From the anthropological perspective, this ignores the contextual factors such as the presence or absence of rules which are put in place locally to regulate the use of the commons. Ignoring the contextual factors and the historical bases of behaviour has led to Hardin's model being applied in an almost universal manner:

> Unfortunately, many of those using the tragedy-of-the-commons model have failed to recognize its assumptions and verify their applicability to the case at hand. . . . The individualistic bias of the commons models leads to underestimates of the ability of people to cooperate in commons situations and contributes to the tendency to avoid social, historical, and institutional analysis.[27]

Market economy analysis has generally undervalued the social restraints that exist at a local level between individuals, and has therefore wrongly branded common property as no one's property or, in other words, as an open access situation. In fact, common property is usually a very distinct recognition of limits on the part of the users of the commons.

But it is not necessarily the goal of this chapter to vindicate common property as a system of ownership, but rather, to illustrate that it is a red herring when it comes to the problems in the fishery. There is every indication that it was not common property that caused havoc in the fishery. Instead, it was powerful technologies operating in an unregulated open access situation and motivated by market economy rationale, which caused the ongoing crises.

THE DISLOCATION OF BLAME

This re-evaluation of the commons in various disciplines becomes important when the model is misused by various practitioners, fishery managers included, because they have failed to understand the nature of the problems of overcapacity and depletion. This is an especially important aspect of the ongoing lack of intercultural understanding in the conservation debate between the Northern and Southern hemispheres. To quote Chopra, Kadekodi, and Murty:

> . . . the concept of common property (like any other property structure) must be defined in a manner that includes the nature of the institutions that enforce it. Not doing this can lead to erroneous policy conclusions in a situation in which these institutions tend to become weak and ineffective. Privatization of common property resources, for example, is often suggested as a solution to 'the tragedy of the commons.' The protagonists of this view base their postulations on (a) the prohibitive cost of reaching and policing agreements to determine rates of use or exploitation of common property, or (b) the interpretation of common property as 'ownership by all' rather than by a group functioning under a specific set of rules.[28]

McCay and Acheson sum up the "erroneous policy conclusions" related to the "common property problem":

> It can be argued that the common property status of resources is neither a necessary nor a sufficient explanation for resource depletion and economic impoverishment. Problems blamed on common property rights, such as depletion of resources and impoverishment of communities, may be more closely related to capitalism and other manifestations of a colonized and industrialized world than to common property per se.[29]

It is obvious that a perspective which sees capitalism and industrialization as the real problem, would offer a very different set of solutions to the difficulties that exist between conservation and development than would one that accepts common property as the culprit.

This kind of re-evaluation of the ideas of common property which I have just outlined, is not evident in the approach resource managers have taken to the Northwest Atlantic fishery up to the declaration of the fishing moratorium in 1992. Although attention shifted from government intervention to privatization as a means of solving the problems in the fishery, there was never any questioning of market economy. Instead, conservation measures continued to be focused on the institution of regulatory measures and property rights. Echoing Chopra's analysis of the standard market economy complaints against common property as quoted above, the *Scotia-Fundy Groundfish Task Force Report* outlines the rationale behind the privatization of shares of the annual fish quota in this way:

> Fisheries management employs public resources to generate private gain. The process should be made as efficient as possible to minimize the cost to Canadian taxpayers. Management has evolved toward a system demanding a high degree of administrative, scientific, and enforcement support while manpower and financial resources have been declining. In this light more efficient management measures must be sought. In addition, the management regime should be designed to encourage fishermen to become more efficient, which should in turn produce more benefits overall.[30]

The received wisdom of a deregulated market economy that pervades this quote, offers little hope that measures such as privatization will limit overexploitation. What these measures do insure is that life will be made easier for government regulators — if any semblance of a fishery survives this latest crisis — because ownership of the fish will gravitate toward large reserves of capital. Therefore, there will be far fewer participants in the industry. Regarding the relationship between privatization and conservation, John Ferguson states:

> While there is no doubt that our society would benefit from restricting access to many more resources in order to adequately deal with situations of unrestricted access (i.e. the east coast fisheries), this does not mean that privatized property is the only or even the most desirable way to do that. "Free market environmentalism" sees private property as nearly a panacea for environmental degradation.[31]

This same assumption was increasingly reflected in fisheries conservation literature as the mandate for a comprehensive national management waned in light of increased costs and repeated failures.

If we analyze the problems of common property by accepting as given and eternal the workings of modern economy, then yes, common property is the problem since it does not fit our modern paradigm based on profit and private property. In other words, the fishery is not capitalist enough. But, if we do not accept the workings of modern economy as given, and if we relate the problems in the fishery to the environmental problems in the world generally, then serious questions have to be raised about our society's approach to our natural surroundings. In short, we have to go beyond concepts of regulation and property rights.

If all the condemnations of common property exploitation in the fishery quoted at the beginning of the chapter were reframed as condemnations of modern economic technological realities, there would be an immediate fusion between the problems that exist in the fishery and the range of other environmental problems. There would then also be linkages and perceptions that would redefine the dynamics of overexploitation. With this perspective in mind, the analysis of "development projects" undertaken in the Southern Hemisphere can shed light on the problems in the Northwest Atlantic fishery, as well as on the more general discussion of the commons. The reason this is a fruitful comparison is that there is a more obvious contrast between the agenda of centralized authorities, such as the World Bank, and the goals of local cultures in the Southern Hemisphere, which are being "developed." The same kind of colonial relationship exists between Ottawa and Atlantic fishing communities. It is just less clear since it occurs within the same country.

There is no question that common property cannot be aligned to market economy, as is repeatedly contended. The difficulty lies in defining the dynamic within this juxtaposition that is anathema to the living diversity of the Earth. Because of the overexploitation and depletion that is dictated by economic realities of exchange, common property remains problematic to the extent that we do not face the fact that viewing our natural surroundings as forms of standing capital from which we can expect an economic return, will promote their destruction.

As is clear in the Northwest Atlantic fishery, the "focus on conservation" — as conceived and developed within the institutional arrangements of modern market economy — was implicated and compromised at every turn in the attempt to limit exploitation. Although the 200-mile limit and the regulatory frameworks that followed were all embarked on in the name of conservation, they became, for all intents and purposes, enclosure movements which promoted industrial production that was increasingly consolidated in the hands of a few.

If there was a "rush to fish" which caused overcapacity and depletion, it was not caused by common property, but by the dis-integration of conservation and development within the socially-impoverished categories of late capitalism. Modern economic and technological imperatives are allowed to be present in the fishery, and it is the defined purpose of conservation to bring these forces to a sudden halt when a level of exploitation is reached. This is an entirely unrealistic expectation in the context of modern economic realities, and is all but facile in its attempt to control deeply entrenched historical forces. Once again, conservation of the natural world fails because it does not make the structures and processes of modern life sufficiently problematic. This is the common problem we face.

NOTES

1   W. E. Block. 1990. *Economics and the Environment: A Reconciliation*.
     Vancouver: The Fraser Institute, p. 289.

2   Arthur McEvoy. 1986. *The Fisherman's Problem: Ecology and Law in the
     California Fishery*. New York: Cambridge University Press, p. 12.

3   E. P. Durrenburger & G. Palsson. 1987. "The Grass Roots and the State:
     Resource Management in Icelandic Fishing." *The Question of the
     Commons*. [Bonnie McCay & James Acheson, eds.]. Tuscon: University of
     Arizona Press, p. 371.

4   Fisheries and Marine Service. 1976. *Policy for Canada's Commercial
     Fisheries*. Ottawa: Department of the Environment, p. 39.

5   Parzival Copes. 1977. *Canada's Atlantic Coast Fisheries: Policy
     Development and the Impact of Extended Jurisdiction*. Burnaby: Simon
     Frazer University, p. 3.

6   Michael Kirby. 1983. *Navigating Troubled Waters: Report for the Task Force
     on the Atlantic Fisheries*. Ottawa: Minister of Supply and Services, p. 32.

7   E. Weeks & L. Mazany. 1983. *The Future of the Atlantic Fisheries*.
     Montreal: Institute for Research on Public Policy, p. 2.

8   Jean Haché. 1989. *Scotia-Fundy Groundfish Task Force Report*. Ottawa:
     Minister of Supply and SErvices, p. 9.

9   Richard Cashin. 1993. *Charting A New Course: Toward The Fishery Of The
     Future*. Ottawa: Minister of Supply and Services, p. 15.

10   L. S. Parsons. 1993. *Management of Marine Fisheries in Canada'*. Ottawa:
     National Research Council, p.118.

11   J. R. Angel, D. L. Burke, R. N. O'Boyle, F. G. Peacock, M. Sinclair, & K. C.
     T. Zwanenburg. 1994. *Report of the Workshop on Scotia-Fundy Groundfish
     Management from 1977 to 1993*. Can. Tech. Rep. Fish. Aquat. Sci. 1979: vi,
     p. 1-2.

12   G. Hardin & J. Baden. 1977. *Managing the Commons*. San Francisco:
     Freeman.

13   Hardin & Baden. (1977:17).

14   Hardin & Baden. (1977:20).

15   Hardin & Baden. (1977:22).

16   Hardin & Baden. (1977:24).

17   Hardin & Baden. (1977:29).

18   R. J. Smith. 1988. "Private Solutions to Conservation Problems." *The
     Theory of Market Failure: A Critical Examination*. [T. Cowan ed.]. Fairfax:
     George Mason University Press, p. 342-3.

19   John Ferguson. 1995. "Privaterianism: Premises, Promises. . : Private
       Property, Collective Action, and Sustainable Development." York
       University: Social & Political Thought, (unpublished article), p.16-17.

20   D. Bromley & M. Cernea. 1989. *The Management of Common Property
       Resources*. Washington: World Bank, p. 6-7.

21   Vandana Shiva. 1989. *Staying Alive: Women, Ecology, and Development*.
       London: Zed Books, p. 88-89.

22   Koula Mellos. 1988. "The Conception of "Reason" in Modern Ecological
       Theory." *Canadian Journal of Political Science*. Vol. XXI, No. 4, December,
       p. 719.

23   Arthur McEvoy. 1987. "Toward an Interactive Theory of Nature and
       Culture: Ecology, Production, and Cognition in the California Fishing
       Industry." *Environmental Review*. Vol: 11, No. 4, pp. 296-7.

24   Shiva. (1989:84).

25   McCay & Acheson. (1987:xiv)

26   McCay & Acheson. (1987:xiii-xiv).

27   McCay & Acheson. (1987:7).

28   K. Chopra, G. Kadekodi, & M. Murty. 1990. *Participatory Development*.
       London: Sage, p. 25.

29   McCay & Acheson. (1987:9).

30   Jean Haché. (1989:10).

31   John Ferguson. (1995:3).

VI

Conclusion: The Political Economy of Depletion and Dependence

Despite the wealth of helpful theory, there have been very few success stories of fisheries management in practice.

Pitcher and Hart[1]

lthough he was not directly concerned with depletion of natural communities, Harold Innis' analysis in *The Cod Fishery: The History of An International Economy*, points to two reasons for this later ecological failure linked to the historical realities in the Northwest Atlantic fishery: 1) the expansion of commodity relations in the context of an increasingly industrialized fishery which put intense pressure on marine communities; and 2) the history of international relations and colonial arrangements left Atlantic Canada in a state of dependence both economically and politically, which led the fishery to have a set of development priorities over which it had minimal control.[2] It was these earlier realities which — during the resource management mandate in the period from 1977 to 1994 — undermined the conservation goals set out in the *Policy for Canada's Commercial Fisheries* on the eve of the declaration of the 200-mile limit.

Since the collapse of marine biotic communities in the early 1970s, and the various regulatory responses to that collapse (including nationalizing the fish grounds), resource managers have been attempting to control exploitation, while at the same time trying to understand the workings of marine biotic communities that have been destabilized by overexploitation. What this has amounted to, finally, is resource management as crisis management. Almost all fishery policy that now exists, has come about from inquiries into breakdowns in the industry. Thus, these policies reflect not the fulfilling of the twin mandates of ecologic and economic stability, but rather, the sacrificing of conservation policy to assuage the cries for more fish. This history of colonial dependence led, in the period immediately following the nationalization of the coastal zone, to the fishing industry being seen as an engine of jobs and economic activity when, in fact, it had been made very vulnerable by the overexploitation of the international fleet.

This national policy of using the fishery as the engine of development (to overcome regional disparity within Canada) appeared most intensely in the expansion of the processing and catching capacity of the larger fish companies, and was subsidized by agencies such as the Federal Department of Regional Economic Expansion. Barrett sums up the history of the relationship between government and large fish companies such as National Sea Products in this way:

> The history of National Sea Products is one of growth and expansion under the protective wing of a developmentist state, especially in the 1970s. In payment for this public tutelage, the company took advantage of every opportunity to exploit underutilized species or new species of fish, and to expand efforts into more traditional fisheries. Centralism, concentration, and technological modernization became its hallmarks. In spite of this seeming orderly expansion, however, anarchy and frenzied overexploitation prevailed. When fish stocks were threatened, the company could only respond by increasing efforts in other areas or by diverting capital out of the fishery or out of the country altogether. To such an organization, conservation and rational management were an anathema.[3]

When the recession of the early 1980s set in, this expansion caused a debt and liquidity crisis in the recently expanded fish companies who had huge stock piles of inventory which they couldn't sell. This crisis led the Federal

Government to set up the Kirby Task Force to inquire into the problems in the fishery. The task force report identified the Canadian Government's conflictual response to fishery issues after the declaration of the 200-mile limit in 1977:

> While the Department of Fisheries and Oceans was slowly tightening up the licensing regime with one hand (and preaching constraint), it was passing out subsidies for fishing vessel construction with the other, as were provincial loan boards.[4]

This drive for economic development led to a situation where, by 1981, the domestic Canadian fleet surpassed the catching capacity of the international fleet which had decimated marine communities in the 1970s. Despite its massive expenditure on regulatory infrastructure, the Canadian Government ended up doing little more than internalizing the very processes of industrial expansion which had destroyed the marine communities in the international context.

Because of the expansion of the Canadian fleet, catch levels increased throughout the early 1980s and leveled off in the mid-1980s before beginning to drop dramatically. Prices paid for fish were at an all time high in the mid-1980s. In other words, despite falling catches, the increased value promoted exploitation of marine communities which were on the brink of collapse. What became clear in the aftermath of collapse is that the fish were always in a vulnerable state, and it was only the increased efficiency and catching capacity of the Canadian fleet which generated increased catches, and was not due to the recovery of the health of marine communities.

In 1989, the *Scotia-Fundy Groundfish Task Force Report* — an inquiry into overcapacity in the groundfish fleet — stated that there was five times the catching capacity in the fleet needed to catch the annual quota. Along with the recognition of overcapacity in the fleet, the report reflects a fundamental change in the Federal Government's approach to the fishery. In contrast to *Policy for Canada's Commercial Fisheries* which saw its mandate in

terms of putting in place a centralized and publicly funded regulatory infrastructure to manage the fishery, and the Kirby Task Force which consolidated
the position of larger companies in the fishery, the 1989 report was more
interested in moving toward the wider government initiative linked to privatization and deregulation of economic activity. A central aspect of this
increased efficiency was the expansion of the Enterprise Allocation program
which had begun with the Kirby Task Force, and which turned the fish in
the ocean into transferable private property — in the form of ownership of a
share of the annual Total Allowable Catch — which was granted to the larger
participants in the industry. This approach assumed that private property
promoted more rational use of the resource, as opposed to the "rush to fish"
impetus which was inherent in the quota system.

As it gradually became clear in the late 1980s that the regulatory
mandate as set out in *Policy for Canada's Commercial Fisheries* had failed, the
Federal Department of Fisheries and Oceans abandoned the mandate of a
comprehensive regulatory infrastructure funded by Canadian taxpayers. By
beginning a program to privatize and deregulate the fishery, the Canadian
Government acknowledged that it had manifestly failed to fulfill the goals of
promoting ecologic and economic stability in Atlantic Canada. In moving
towards privatization, the Canadian government was promoting the global
processes which had depleted biotic communities, and at the same time,
increasing the vulnerability of Atlantic coastal communities.

As I have discussed throughout the book, the processes of development leading to ecological collapse preceded conservation initiatives. This
reversal resulted in a scenario whereby the goals of conservation were not ful-
filled. Whereas resource conservation and sustainability rest on a clear recognition of the generative capacities of ecological processes, and the setting of
exploitation levels which do not exceed that capacity, when development precedes conservation, it becomes very difficult to gain a clear understanding of
these ecological processes because they have been disrupted by ecological collapse, and existing exploitation levels are already in place and resist limits.

The futility of the "wealth of helpful theory" referred to above by
Pitcher and Hart, are the resource management perspectives under which the

Federal Department of Fisheries and Oceans operated. What rendered them futile is their implicit acceptance of the processes of modern economy which had promoted the depletion of marine communities and dependence in Atlantic human communities. As Barbara Neis states, it is the failure to take natural barriers into account which "contributed to the crisis in Fordism in the fishery and these have continued to hamper efforts to establish a new effective regime of accumulation, not only in the North Atlantic, but globally as well."[5] Although forever mired in an ongoing economic and ecologic crisis, fishery managers nonetheless could not or would not acknowledge the "financial vortex" which undermined any conservation initiatives. Instead, managers saw the problems in the fishery in terms of its "poor fit" into modern economic categories.

What the "race for quota" is really referring to is not the common property problem, but the location of conservation — as it operates in a quota system — within modern political economy. Conservation is not an on/off switch for destructive behaviour imposed by an external authority at some upper level of exploitation at the last minute. In other words, to allow all the workings of modern economy to operate according to technology and economic pressure, and then to expect all this to grind to a halt when catch levels are reached, is the analytical equivalent of solving waste management problems by standing at the gate of the landfill site with a whistle. Therefore, the implicit acceptance of the workings of modern economy by resource management and sustainability perspectives, combined with the fact that these perspectives generally initiate conservation in the aftermath of crisis, limits their effectiveness in promoting conservation of the natural world.

This resistance to useful analysis is reflected again in the *Report of the Workshop on Scotia-Fundy Groundfish Management from 1977 to 1993*. Presented in the report is a table which conveys over 300 instances where discarding, dumping, misreporting, and highgrading of catches — activities which are promoted by the "last minute" location of the quota system as exploiters attempt to both meet quota requirements and maximize economic return — were seen as the central problem in gathering reliable data for the purposes of developing a groundfish management plan. The report then goes on to state that:

> It has been interpreted that changes in environmental conditions [read: colder water] have been a major contributor to the declines in northern cod off Newfoundland and Labrador. By inference, it has been concluded by some that environmental conditions have increased natural mortality in other areas where stocks have also declined steadily since the late 1980s. The continuous growth of the gray seal populations since the extension of jurisdiction has also been considered to be important.[6]

Once again, although the information on dumping and misreporting is included within the text of the workshop report, along with the discussion of seals and cold water, the abstract makes no mention of misreporting and dumping, but states that "Two papers. . . evaluated the degree to which changes in natural mortality (by, respectively, environmental trends and seal predation) over time have compromised our ability to attain management objectives."[7] This kind of deflection of analysis away from a discussion of the relationship between economic development and environmental problems, is an indication of the extent to which resource management and sustainability, almost by definition, take the assumptions of modern economy for granted. This reality limits their effectiveness in both controlling exploitation and in analyzing failures in the aftermath of overexploitation.

Although never directly concerned with the ecological collapse of natural communities, Harold Innis' political economy analysis of the cod fishery in the 1930s is far more able to make a contribution to the political economy of depletion and dependence in the 1990s than is resource management fisheries literature, or is global sustainability literature which shares many of the same assumptions as fisheries literature. Rosemary Ommer makes a similar point in her book *From Outpost to Outport: A Structural Analysis of the Jersey-Gaspe Cod Fishery, 1767-1886*:

. . . the roots of the current major issues in the inshore fishery, and of some of the problems of eastern Canadian regional underdevelopment, lie in the way the early merchant fishery was first established and conducted. . .[8]

By recognizing the significance of the relationship between natural processes and economic structures, and the relationships which accompany various forms of production in historical terms, political economy analysis makes a contribution to the current problems in the fishery. Of the dislocation of Atlantic Canadian communities, Innis expressed this concern in the midst of economic changes in the 1930s:

Nova Scotia turned to the interior in Canada and the United States, and retreated from world markets where she found herself in competition with the capacity of large-scale fish production in other important countries. The results of the retreat were evident in the revolution from an economy facing the sea with a large number of ports to an economy dependent on a central port [Halifax] and railways to the interior. . . . The disappearance of an active commercial region as a result of the impact of machine industry has been a major calamity to the fishing regions of France, New England, Nova Scotia, and Newfoundland. . . . The transition from dependence on a maritime economy to dependence on a commercial economy has been, slow, painful, and disastrous.[9]

In comparison to the slow, painful, and disastrous transition of the 1930s, it is hard to imagine the words needed to describe the transition in the 1990s for these same communities who are struggling in the aftermath of ecologic collapse.

If the political economy of depletion and dependence is seen as being a significant perspective in understanding the destruction of marine biotic communities in the Northwest Atlantic, this scenario becomes positively daunting when it is applied against current initiatives directed toward economic globalization. The marginalization of peripheries such as the Atlantic coastal communities, by centres such as London and Ottawa, and their consequent dependence, is exacerbated by the "level playing field" of globalization perspectives in which there is, increasingly, peripheralization of the centre as the economic forces intensify, and any mandate other than that which suits the community of capital, becomes an inefficiency and irrationality on the level playing field of privatization, deregulation, and free trade.

From the perspective of Atlantic coastal communities, and local cultures all over the globe, the economic forces instituted by interests from the centre exploited the natural processes on which they depend to the point of collapse. This leaves these communities vulnerable to forces over which they have no control. With the shriveling mandate to implement remedial support by these interests, such as unemployment insurance in the Canadian context, these communities have nothing with which to bargain as economic interests look elsewhere for profit. It is this state of vulnerability and marginalization which is sure to expand and generate social and political strife in the future. The "focus on conservation" as represented in resource management and sustainability initiatives, have done little to mitigate this process and in fact, represented a "struggle for resources carried on by other means."

EXTERNALIZING INTERNALITIES:
COMMUNITY-BASED RESISTANCE TO GLOBALIZATION

A modern theory of resource management should recognize
that relations between humans and their environments are
complex, involving many aspects of sociocultural systems,
and that a definition of resource management that stresses

rationality with the coercive backing of the state as the only
kind of management is shallow indeed.

Anthony Stocks[10]

Broadly speaking, there are two main approaches to environmental
problems related to pollution and depletion. The first could be regarded as a
strategic, instrumental, and technical approach. It is usually science and planning oriented, and does not call the modern human project related to science, technology, and capitalism into question. Instead, it focuses on monitoring activities and reforming certain practices which are deemed to have
negative consequences for the environment. This could be regarded as a
problem-solving approach.

The other approach to environmental issues is based on social and
cultural analysis, and begins with the recognition that the structures of
everyday life and the structures that cause environmental problems, are one
and the same. Therefore, it is all but impossible to separate what is causing
environmental problems from the texture of a whole society. Discussions
from this perspective relate to the hegemonic domination of current realities and reject the concept that, since environmental problems are connect-
ed with the fabric of everyday life, that it is possible to develop and maintain a regulatory perspective which can mitigate problems, since regulation
itself can be part of the problem. In other words, dealing with environmental issues may require more than solving problems, it may be necessary to
solve history by making modern everyday life problematic in order to deal
with environmental problems. If these two perspectives of solving problems
and solving history are seen as poles on a continuum, there is an inverse
relation between the extent to which current modern realities are universalized and strategized on the one end, and how much moral, ethical, or gen-
der-political economy considerations are taken into account in the analysis
of environmental problems on the other.

A significant aspect of the strategic, instrumental approach could be
defined in terms of internalizing within hitherto narrow, short-term economic considerations, what had been external to these considerations in terms of

pollution or depletion. In order to internalize externalities and promote long-term mitigation of problems, it is necessary to find ways to value non-economic realities such as natural processes, so that they can become variables that can be taken into account in economic terms. Within resource management and sustainability perspectives then, conservation involves valuing the non-economic, both with regard to human preference, as well as in terms of natural processes. As pointed out above, this does not involve challenging the modern human project, but rather, it is a problem solving strategy related to the reforming of certain of its practices.

But from the perspective of the political economy of depletion and dependence, the ramifications of internalizing externalities for Atlantic Canadian coastal communities, can be seen in terms of the intensification of marginalization and peripheralization. Internalizing externalities involves universalizing modern economic realities by converting what had been external to those equations into values which are comprehensible to them. After the world is homogenized into values that suit modern economy, these hitherto externalities are then turned over to those who are destroying them. As I have argued, the "focus on conservation" linked to resource management and sustainability, became the very processes by which exploitation was intensified in the Northwest Atlantic. In other words, conservation became an enclosure movement which served the interests of modern economy.

Initiatives to conserve the natural world require a viable social context in which to manifest themselves. In turn, economic globalization is not a social context that offers any hope for conservation. This is the "intellectual ruin" called development, to which Wolfgang Sachs refers. If there is to be any future for the conservation of biotic communities, what is therefore required is the creation of a viable social context which can only begin by resisting the forces of economic globalization. A more viable approach to conservation then, — and this locates me within social and cultural analysis — involves an approach to environmental problems based on externalizing internalities. Rather than universalizing modern economy — as the internalization of externalities does by converting the world into its categories of understanding — externalizing internalities begins by making modern economy problematic. This process involves extricating nature and humans from categories of understanding linked to modern economy, and focuses on the larger project of defining a viable social basis for the discussion of conservation which is not implicated in the current ruin called development. For those people concerned with environmental issues related to the conservation of the natural world, the basis for the social and cultural project related to externalizing internalities, has a polar opposite starting point to the reform environmental project of internalizing externalities into the modern economic equation. It is this social and cultural project which should form the primary basis of the environmental debate, rather than the reform of the traditional economic model.

In this context, conservation begins in the implicit acceptance of one's place in human community and natural community. In large part then, conservation is a social and cultural issue, not a regulatory problem. Like Aboriginal people in Canada, who in full recognition of their cultural difference, have set about reclaiming their sense of community from the Canadian Department of Indian Affairs, or like local communities in the Southern hemisphere who struggle against the edicts of international financial institutions, Atlantic coastal communities can set about initiating a "sea claim" to reclaim control of nearby natural communities on which they have depended, based upon a radically democratic conception of interrelationships. In contrast to the regulatory basis of Canada's declaration of the 200-mile limit which is based on the recognition of predatory international fishing practices — that were then only replicated internally within national boundaries — coastal communities require a more viable social and cultural basis for challenging the forces of depletion and dependence.

There is an interesting contrast that I experience living in an Atlantic coastal fishing community. On the one hand, there is little or no regulation of law and order in the community itself. In general, the Royal Canadian Mounted Police cruiser car rolls through our community once a month and doesn't even bother to slow down. If anything goes wrong or something goes missing, people usually have a good idea who is responsible, and the issue is dealt with in short order. It is this lack of regulation in the social context of community life which is in such stark contrast to the highly-regulated, divisive, strife-ridden, and competitive fishing activity which is engaged in by these same people. What underwrites this difference is the presence of modern economic and technological realities in the fishery, and the behaviour which accompanies them. Present also in this contrast is the comparison between what I have conveyed in terms of a viable social context for conservation and the fact that modern economic and technological realities — and their accompanying institutional arrangements — are not a viable social context for conservation.

In setting up this contrast in social contexts, and linking conservation to a social context generally, I put myself in the position of having to solve history in order to solve problems in the world's ocean fishery, especially when it is considered that, at present, history is moving most concertedly in the direction which makes the conservation of the natural world increasingly difficult.

Resistance to the colonizing forces of modern economy have appeared in various parts of the world. The resistance of the Kanak people to French colonization in what is known as New Caledonia in the South Pacific, is especially noteworthy. In a letter to the Kanak entitled "The Policy of the Severed Flower" in which he discusses their liberation document, Dominique Temple outlines what he perceives as the contrast between Kanak culture and the one based on modern economic production: ". . . the principle according to which power is in proportion to giving [in Kanak culture], which is the inverse of Western society's principle, where in power is in proportion to accumulation."[11] Temple relates the centrality of gifts to societies based on reciprocity and redistribution, or what Karl Polanyi described as embedded social relations, as opposed to the disembedded relations of modern economy.[12] In the context of the circulation of goods as a form of social exchange, reciprocity and redistribution in the Kanak culture has nothing to do with the accumulation of goods. Rather, circulation is defined by giving, not taking.

The question Temple asks, and this is central to the project of externalizing internalities, is this:

> . . . if the categories of the economy of reciprocity can be
> interpreted as categories of the exchange economy, then one
> system is reducible to the other and one can integrate the
> Kanak system into the economic system that Western society
> is trying to impose on the whole world.[13]

This reducibility and integration is central to the reform project of internalizing the externalities related to valuing the non-economic, and thereby universalizing the perspective of modern economy and erasing the specificity of the history of modern economy. As Temple states:

> For if one could. . . interpret the principles of the Kanak
> economy [in Western terms]. . . without fundamentally
> changing their nature. . . the Kanak would have no choice
> but to adapt to the Western world and be integrated into the
> exchange system, hoping that this integration be worked out
> in such a way as to protect their well understood interests.[14]

But, "if the answer to this question is in the negative," then this integration spells "suicide" for Kanak culture whereby

> the very notion of value in the system of reciprocity is dis-
> qualified. . . by replacing the gift and reciprocity with
> barter, one signals the rapid establishment of economic
> exchange, since barter can be considered as the forerunner
> of exchange. The roots of Kanak values are replaced with
> those of Western values.[15]

For Temple, this integration leads to the erasure of the basis of Kanak reciprocity and masks the fact that one system is not reducible to the other, and that these systems are, in fact, "antagonistic and do not even beget the same concepts, the same ideologies, the same notion of value. . . and are two different systems of civilization."[16] Central to this difference is the presence of the sacred within reciprocity. This is not the sacredness of a transcendent god, but is a sacredness that is "rigorously and specifically human."[17] In the context of discussions of conservation, this animated sense of being can be more broadly located in all sentient beings, not just in humans. This manifoldness then links human communities with the whole of the life world.

This may seem like quite a leap from the colonial history of the Northwest Atlantic fishery to a discussion of Kanak resistance to French colonization in the South Pacific. What I am trying to point out is that the first step toward a viable discussion about conserving the natural world, requires a challenge to the universalizing categories of modern economy. This is precisely what Temple is doing in his discussion of the Kanak and the French colonizers. By externalizing the internalizing tendencies of modern economy, the Kanak base their resistance on a conception of themselves which is of an entirely different order to modern economy. Extricating human communities and natural communities from this production model is central to the conservation project.

In the essay *Indigenous Law and the Sea* by Moana Jackson, a Maori activist from New Zealand, there is a similar perspective on the relationship between local community and increasingly global economic relations:

> The doctrine [of freedom of the seas] framed to protect seventeenth-century mercantilism thus continues to protect the same interests today. The interests of smaller states, or of indigenous Pacific peoples living under colonial regimes, are scorned. . . . The freedom to use the seas has become a license to abuse, with a consequent depletion of the fisheries. . . . [A]ny attempt to remedy the environmental consequences must also seek. . . to restor[e] the rights of those indigenous people who have been marginalized and disrupted by its exploitation. Such reform cannot be achieved through interna-

tional legal legerdemain or by a mere refinement of
existing conventions. It can be done only by addressing
the fundamental reasons for the exploitation and the
complimentary rationale of the law that permitted it.[18]

It was the "existing conventions" which suppressed indigenous values and
promoted the abuse of the oceans. For Jackson then, they hold little hope of
promoting the conservation of the natural world. Instead, Jackson sets forth
a law of the sea "deeply rooted in Maori cultural values," and linked to the
idea that the sea is a "gift" which human communities may use "in a way that
will sustain its bounty."[19]

In his analysis of the regulatory failure of the Northwest Atlantic fishery in *Controlling Common Property: Regulating Canada's East Coast Fishery*,
David Ralph Matthews critiques the regulatory frameworks used by the
Canadian Department of Fisheries and Oceans, and their attempt to rid the
fishery of common property. Like Temple and Jackson, Matthews challenges
dominant perspectives which universalize modern economy. He sees common property and its connection to community as a possible solution to regulation in the fishery:

. . . the link between commons and community is not only
structural, but also symbolic. . . . [I]t is clear that scholarly
attention needs to be directed to the social-psychological
dimensions of such reconstructions of meaning. That
analysis should focus on the identification with and commitment to community. . .[20]

This is the project I would identify in terms of externalizing internalities in
the name of conserving human and natural communities.

To hard-headed types well-versed in the realities of the modern
"real world," this kind of discussion about solutions to the crisis in the
ocean fishery may seem very impractical. I say to them: What could be
more impractical than putting 50,000 people out of work because of the

ecologic collapse of the North Atlantic Cod? What could be more impractical than a sinking ship? What could be more impractical than rattling
around inside an intellectual ruin called economic development, working the
levers that only bring on disaster?

It may be that solutions to environmental problems, such as the
destruction of the natural world, are finally of a different order than what
caused the problems. This is the "social-psychological" initiative related to
the recognition that conservation can only begin in the implicit recognition
of animated membership in human and natural communities. This initiative
cannot be imposed from outside, but requires a located sense of belonging
and membership which can both resist a global economy, and at the same
time insure a collective and radically democratic recognition of those located
interests. This project has a far greater chance for success if it is initiated
before ecologic collapse rather than struggling against the forces of exploitation and dislocation in the aftermath of collapse.

POSTPONING THE DAY

And these dismaying repetitions — this unconscious limiting
or coercion of the repertoire of lives and life-stories — create
the illusion of time having stopped. In our repetitions we
seem to be staying away from the future, keeping it at bay.
What we call symptoms are these (failed) attempts at closure,
at calling a halt to something. Like provisional deaths, they
are spurious forms of mastery.

Adam Phillips[21]

What is clear here is that the ongoing crises in the Northwest
Atlantic fishery are not just fisheries problems, but are representative of modern economic relations generally. In this sense there can be no fisheries solution that does not, at the same time, address these wider realities of "normal
business." In fact, what is peculiar about the fishery is not the excessiveness

of the exploitation that occurs there, as compared to other sectors of modern economy which are not so obviously in a state of ongoing crisis. What is peculiar about the fishery is that it is made up of human communities who are dependent on immediately adjacent natural communities, and have no alternate way of making a living. That is what is unusual about the fishery. The levels of exploitation which support life in Ontario, Missouri, or Switzerland are no more viable. It is just that increasingly globalized networks of resource acquisition and market distribution, mask this failure in viability. It is the forests of Indonesia or the fish from the Western Pacific which create raw materials for the "developed" centre. Imagine all the fishers of Newfoundland jumping into their boats and sailing off to Micronesia to catch fish to support life back home, and you have a comparable image for the network that serves life in industrialized societies in general. This masking of failure in viability related to exploitation levels will come to an end, and when it does, it will be global in scope. Instead of having a globalized economy, we will face the possibility of increasingly globalized environmental wars. All for the sake of making a dead thing grow.

NOTES

1   T. Pitcher & P. Hart. 1982. *Fisheries Ecology*. London: Croom Helm, p. 344.

2   Harold Innis. 1940. *The Cod Fishery: A History of An International Economy*. Toronto: University of Toronto Press.

3   Gene Barrett. 1984. "Capital and the State in Atlantic Canada: The Structural Context of Fishery Policy Between 1939 and 1977." *Atlantic Fisheries and Coastal Communities: Fisheries Decision-Making Case Studies*. [Cynthia Lamson and Arthur Hanson eds.] Halifax: Dalhousie Ocean Studies Programme, p, 96.

4   Kirby (1983:20).

5   Barbara Neiss. (1993:102).

6   Angel (1994:115).

7   Angel (1994:1).

8   Rosemary Ommer. 1991. *From Outpost to Outport: A Structural Analysis of the Jersey-Gaspe Cod Fishery, 1767-1886*. Montreal & Kingston: McGill-Queen's University Press, p. 3.

9   Innis (1954:507-508).

10 Anthony Stocks. 1987. "Resource Management in an Amazon *Varzea* Lake Ecosystem: The Cocamilla Case." *The Question of the Commons*. [Bonnie McCay & James Acheson eds.] Tuscon: University of Arizona Press, p. 110.

11 Dominique Temple. 1988. "The Policy of the Severed Flower." *INTERculture*, Vol. 21, No. 1, Issue 98, p. 12.

12 Karl Polanyi. 1957. *The Great Transformation*. Boston: Beacon Press.

13 Temple (1988:14).

14 Temple (1988:14).

15 Temple (1988:16).

16 Temple (1988:17).

17 Temple (1988:24).

18 Moana Jackson. 1993. "Indigenous Law and the Sea." *Freedom for the Seas in the 21st Century: Ocean Governance and Environmental Harmony*. [Jon Van Dyke, Durwood Zaelke, & Grant Hewison eds.] Washington: Island Press, p. 43.

19 Jackson. (1993:46).

20 David Ralph Matthews. 1993. *Controlling Common Property: Regulating Canada's East Coast Fishery*. Toronto: University of Toronto Press, p. 94.

21 Adam Phillips. *London Review of Books*, Feb. 10, 1994:13.

BIBLIOGRAPHY

Angel, J. R., DL. Burke, R. N. O'Boyle, F. G. Peacock, M. Sinclair, and K. C. T. Zwanenburg. 1994. *Report of the Workshop of Scotia-Fundy Groundfish Management from 1977 to 1993*. Can. Tech. Rep. Fish. Aquat. Sci. 1979: vi + 175 p.

Barrett, L. G. 1984. "Capital and the State in Atlantic Canada: The Structural Context of Fisheries Policy between 1939 and 1977." *Atlantic Fisheries and Coastal Communities: Fisheries Decision-Making Case Studies*. [C. Lamson & A. Hanson eds.] Halifax: Dalhousie Ocean Studies Programme.

Block, W. E. 1990. *Economics and the Environment: A Reconciliation*. Vancouver: The Fraser Institute.

Bromley, D. and M. Cernea 1989. *The Management of Common Property Resources*. Washington: World Bank.

Brown, Lester. 1995. *The State of the World*. New York: Norton.

Caddy, J. F. and G. D. Sharp. 1986. *An Ecological Framework for Marine Fishery Investigations*. Rome: FAO, No. 283.

Campana, S. and J. Hamel. 1989. "Status of the 4X cod stock in 1988." CAFSAC Res. Doc. 89/30.

Cashin, Richard. 1993. *Charting a New Course: Toward the Fishery of the Future*. Ottawa: Minister of Supply and Services.

Chopra, K., G. Kadekodi and M. Murty. 1990. *Participatory Development*. London: Sage.

Clark, C. 1990. *Mathematical Bioeconomics*. New York: John Wiley and Sons.

Chatterjee, Pratrap and Mattius Finger. 1994. *The Earth Brokers*. New York: Routledge.

Christy, F. and A. Scott. 1965. *The Common Wealth in Ocean Fisheries*. Baltimore: The Johns Hopkins Press.

Copes, P. 1977. *Canada's Atlantic Coast Fisheries: Policy Development and the Impact of Extended Jurisdiction*. Burnaby: Simon Fraser University.

Copes, P. 1978. *Rational Resource Management and Institutional Constraints: The Case of the Fishery*. Burnaby: Simon Fraser University.

Coull, James R. 1993. *World Fisheries Resources*. London: Routledge.

Cushing, D. H. 1987. *The Provident Sea*. New York: Cambridge University Press.

Department of Fisheries and Oceans. 1979. *Resource Prospects for Canada's Atlantic Fisheries 1980-1985*. Ottawa: Department of Fisheries and Oceans.

Department of Fisheries and Oceans. 1984. *Resource Prospects for Canada's Atlantic Fisheries 1985-1989*. Ottawa: Department of Fisheries and Oceans.

Department of Fisheries and Oceans. 1988. *Resource Prospects for Canada's Atlantic Fisheries 1989-1993*. Ottawa: Department of Fisheries and Oceans.

Department of Fisheries and Oceans. 1989. *DFO Factbook*. Ottawa: Minister of Supply and Services.

Doubleday, W. G. and D. Rivard. 1981. *Bottom Trawl Surveys*. Ottawa: Department of Fisheries and Oceans.

Durrenberger, E. P. and G. Palsson. 1987. "The Grass Roots and the State: Resource Management in Icelandic Fishing." in *The Question of the Commons*. [B. McCay & J. Acheson eds.] Tuscon: University of Arizona Press.

Esteva, G. 1990. *The New Commons*. Canadian Broadcasting Corporation, Ideas: "The Informal Economy." [D. Cayley ed.] November 28, 1990.

Evernden, N. 1985. *The Natural Alien*. Toronto: University of Toronto Press.

Ferguson, John. 1995. "Privaterianism: Premises, Promises. . : Private Property, Collective Action, and Sustainable Development." York University: Social & Political Thought, (unpublished article).

Finlayson, A. C. 1994. *Fishing For Truth: A Sociological Analysis of Northern Cod Stock Assessment from 1977 to 1990*. St. John's: Institute of Social and Economic Research.

Fletcher, H. F. 1977. *Toward a Relevant Science: Fisheries and Aquatic Scientific Resource Needs in Canada*. Ottawa: Ministry of Supply and Services.

Fisheries and Marine Service. 1976. *Policy for Canada's Commercial Fisheries*. Ottawa: Department of the Environment.

Gascon, D. 1988. "Catch Projections." *Collected Papers on Stock Assessment Methods*. [D. Rivard ed.] CAFSAC Res. Doc. 88/61 135-146.

Gavaris, S. 1988. "Abundance Indices from Commercial Fishing." *Collected Papers on Stock Assessment Methods*. [D. Rivard ed.] CAFSAC Res. Doc. 88/61 3-15.

Gavaris, S. 1988. "An adaptive framework for the estimation of population size." CAFSAC Res. Doc. 88/29.

Gavaris, S. 1980. "Use of a multiplicative model to estimate catch rate and effort from commercial data." *Canadian Journal of Fisheries and Aquatic Sciences*. Vol. 37. No.12.

Gray, D. F. 1979. "4VsW: Background to the 1979 assessment." CAFSAC Res. Doc. 79/30.

Gordon, H. S. 1954. "The Economic Theory of the Common Property Resource: The Fishery." *Journal of Political Economy*. Vol. 62. p. 124-42.

Gulland, J. A. and Boerema. 1973. "Scientific Advice on Catch Levels." *Fishery Bulletin* No. 71.

Gunther, P. E. and J. R. Winter. 1978. *Fisheries Rationalization, Employment and Regional Economic Policy*. Wolfville: Acadia University.

Haché, J.- E. 1989. *Scotia-Fundy Groundfish Task Force*. Ottawa: Minister of Supply and Services.

Hardin, G. and J. Baden. 1977. *Managing the Commons*. San

Francisco: Freeman.

Harris, L. 1989. *Independent Review of the State of the Northern Cod Stock*. Prepared for Department of Fisheries and Oceans May 15.

Heilbroner, Robert, 1985. *The Nature and the Logic of Capitalism*. New York : Norton.

Hood, P. R., Macdonald, R. D. S. and Carpentier, G. 1980. *Atlantic Coast Groundfish Trawler Study*. Ottawa: Department of Fisheries and Oceans.

Hutchings, J. A. & Ranson A. Myers. 1994. "What Can Be Learned from the Collapse of a Renewable Resource?: Atlantic Cod, *Gadus morhua*, of Newfoundland and Labrador. St. John's: Science Branch, Department of Fisheries and Oceans.

Innis, H. 1940 *The Cod Fisheries: The History of a International Economy*. Toronto: University of Toronto Press.

International Commission for the Northwest Atlantic Fisheries. *Annual Statistical Bulletins*. Vols. 4- 26. 1954-1976.

Jackson, Moana. 1993. "Indigenous Law and the Sea." *Freedom for the Seas in the 21st Century: Ocean Governance and Environmental Harmony*. [Jon Van Dyke, Durwood Zaekle, & Grant Hewison eds.]. Washington: Island Press.

Johnston, K. 1972. *The Vanishing Harvest*. Montreal: The Montreal Star.

Kirby, M. J. L. 1983. *Navigating Troubled Waters: Report for the Task*

Force on the Atlantic fisheries. Ottawa: Minister of
Supply and Services.

Koeller, P. A. 1980. "Biomass estimates from Canadian research ves
sel surveys on the Scotian Shelf and in the Gulf of St.
Lawrence from 1970-79." CAFSAC Res. Doc. 80/18.

Lamson, C. and A. Hanson. [eds.] 1984. *Atlantic Fisheries and Coastal
Communities: Fisheries Decision-Making Case Studies.* Halifax:
Dalhousie Ocean Studies Programme.

Law of the Sea Discussion Paper. 1974. Ottawa: Department of
External Affairs.

Livingston, John. 1981. *The Fallacy of Wildlife Conservation.* Toronto:
McClelland and Stewart.

Livingston, John. 1994. *Rogue Primate: An Exploration of Human
Domestication.* Toronto: Key Porter Books.

Lusigi, W. 1988. "The New Resource Manager." In *For the
Conservation of the Earth.* Golden: International Wilderness
Leadership Foundation. p.42-51.

Macdonald, R. D. S. 1984. "Canadian Fisheries Policy and the
Development of Atlantic Coast Groundfisheries
Management." In *Atlantic Fisheries and Coastal
Communities: Fisheries Decision-Making Case Studies.* [C.
Lamson & A. Hanson eds.] Halifax: Dalhousie Ocean
Studies Programme.

Marx, Karl. 1959. *Capital.* Moscow: Foreign Language
Publishing.

Matthews, David Ralph. 1993. *Controlling Common Property: Regulating Canada's East Coast Fishery*. Toronto: University of Toronto Press.

Mellos, K. 1988. "The Conception of "Reason" in Modern Ecological Theory." *Canadian Journal of Political Science*. December. p. 715-727.

Mercer, M. C. 1982. *Multispecies Approaches to Fisheries Management Advice*. Ottawa: Department of Fisheries and Oceans.

McCay, B. and J. Acheson. 1987. *The Question of the Commons*. Tucson: University of Arizona Press.

McEvoy, Arthur. 1986. *The Fisherman's Problem.: Ecology and the Law in the California Fisheries, 1850-1980*. New York: Cambridge University Press.

McEvoy, A. 1987. "Toward an Interactive Theory of Nature and Culture: Ecology, Production, and Cognition in the California Fishing Industry." *Environmental Review*. Vol. 11, No. 4. p. 289-305

Mitchell, B. 1989. *Geography and Resource Analysis*. London: Longman. (2nd edition).

Mohn, R. K. 1988. "Yield per Recruit Analysis." *Collected Papers on Stock Assessment Methods*. (D. Rivard ed.) CAFSAC Res. Doc. 88/61 147-167.

Mowat, F. 1990. quoted in "Hibernia Blues" by G. Wheeler. *Now Magazine*. Vol. 10, No. 4, Sept. 27- Oct. 3.

Neiss, Barbara. 1993. "Flexible Specialization: What's That Got To Do with the Price of Fish?" *Production, Space, Identity: Political Economy Faces the 21st Century.* [Jane Jenson, Rianne Mahon, & Manfred Bienefeld eds.]. Toronto: Canadian Scholars Press Inc.

O'Boyle, R. N., J. Simon and K. Frank. 1988. "An evaluation of the population dynamics of 4X haddock during 1962-88 with yield projected to 1989." CAFSAC Res. Doc. 88/72.

Ommer, Rosemary. 1991. *From Outpost to Outport: A Structural Analysis of the Jersey-Gaspe Cod Fishery, 1767-1886.* Montreal & Kingston: McGill-Queen's University Press.

Parsons, L. S. 1993. *Management of Marine Fisheries in Canada.* Ottawa: National Research Council.

Patton, D. J. 1981. *Industrial Development and the Atlantic Fisheries.* Toronto: James Lorimer and Company.

Pearse, P. H. 1982. *Turning the Tide: A New Policy for Canada's Pacific Fisheries.* Ottawa: Governor General of Canada.

Pepper, D. A. 1978. *Men, Boats and Fish in the Northwest Atlantic Cardiff:* Department of Maritime Studies, University of Wales.

Phillips, Adam. 1994. *London Review of Books,* Feb. 10.

Pielou, E. C. 1975. *Ecological Diversity.* New York: John Wiley and Sons.

Pinkerton, E. [ed.] 1989. *Co-operative Management of Local Fisheries*. Vancouver: University of British Columbia Press.

Pitcher, T. J. and P. Hart. 1982. *Fisheries Ecology*. London: Croom Helm.

Polanyi, Karl. 1957. *The Great Transformation*. Boston: Beacon Press.

Polanyi, Karl. 1968. *Primitive, Archaic, and Modern Economies*. [George Dalton ed.] New York: Doubleday Anchor.

Pross, A. P. & S. McCorquodale 1987. *Economic Resurgence and the Constitutional Agenda: The Case of the East Coast Fisheries*. Kingston: Queen's University.

Rivard, D. 1988. *Collected Papers on Stock Assessment Methods*. CAFSAC Res. Doc. 88/61.

Rogers, Raymond A. 1994. *Nature and the Crisis of Modernity: A Critique of Contemporary Discourse on Managing the Earth*. Montreal: Black Rose Books

Rothschild, B. 1983. *Global Fisheries: Perspectives for the 1980s*. New York: Springer-Verlag.

Sachs, Wolfgang [ed.]. 1992. *The Development Dictionary*. London: Zed Books.

Sachs, Wolfgang [ed.]. 1993. *Global Ecology*. London: Zed Books.

Schnute, J. 1985. "A General Theory for Fishery Modeling." *Resource Management*. [M. Mangel ed.] Berlin: Springer-Verlag. 1-28.

Shiva, V. 1989. *Staying Alive*. London: Zed Books.

Sinclair, M., O'Boyle, R., and Iles, T. D. 1981. "Consideration of the stable age distribution assumption in 'analytical' yield mod els." CAFSAC Res. Doc. 81/11.

Sinclair, P. R. 1985. *The State Goes Fishing: The Emergence of Public Ownership in the Newfoundland Fishery*. St. John's: Memorial University.

Sinclair,W. F. 1977. *Management Alternatives and Strategic Planning for Canada's Fisheries*. Ottawa: Fisheries and Marine Services.

Smith, R. J. 1988. "Private Solutions to Conservation Problems." *The Theory of Market Failure: A Critical Examination*. [T. Cowan ed.]. Fairfax: George Mason University Press.

Smith, S. J. 1988. "Abundance Indices from Research Survey Data." *Collected Papers on Stock Assessment Methods*. [D. Rivard ed.] CAFSAC Res. Doc. 88/61 16-43.

Stocks, A. 1987. "Resource Management in an Amazon Varzea Lake Ecosystem: The Cocamilla Case." In *The Question of the Commons*. [B. McCay and J. Acheson eds.] Tuscon: University of Arizona Press.

Task Group on Newfoundland Inshore Fisheries. 1987. "A study of trends of cod stocks off Newfoundland and factors influenc ing their abundance and availability to the inshore fishery." Ottawa: Department of Fisheries and Oceans.

Taussig, Michael. 1980. *The Devil and Commodity Fetishism in South America*. Chapel Hill: University of North Carolina Press.

Temple, Dominique. 1988. "The Policy of the Severed Flower." *INTERculture*, Vol. 21, No. 1. Issue 98.

Troadec, J.- P. 1983. *Introduction to Fisheries Management.* Rome: FAO, No.224.

Weber, Peter. 1993. *Abandoned Seas: Reversing the Decline of the Oceans.* Washington: Worldwatch Paper 116.

Weber, Peter. 1995. "Protecting Oceanic Fisheries and Jobs." *State of the World 1995.* [Lester Brown ed.] New York: Norton.

Weeks, E. & L. Mazany 1983. *The Future of the Atlantic Fisheries.* Montreal: Institute for Research on Public Policy.

Winters, G. H. 1988. "The Development and Utility of Sequential Population Analysis in Stock Assessments." *Collected Papers on Stock Assessment Methods.* [D. Rivard ed.] CAFSAC Res. Doc. 88/61 111-134.

Winters, J. R. 1981. *The Economics and Management of Canada's East Coast Fisheries.* Ottawa: Department of Fisheries and Oceans

Williams, Raymond. 1980. *Problems in Materialism and Culture.* New York: Verso.

World Commission on Environment and Development. 1987. *Our Common Future.* New York: Oxford University Press.

World Conservation Strategy. 1980 Gland: UNESCO, FAO, & IUCN

World Conservation Union, United Nations Environment

Programme, and World Wide Fund for Nature. 1991. *Caring for the Earth*. Gland.

Wooster, W. S. (ed.) 1988. *Fishery Science and Management*. New York: Springer-Verlag.

INDEX

THE NEW RESOURCE WARS

Native Struggles Against Multinational Corporations

Al Gedicks

Aboriginal and environmental coalitions fighting against corporate greed and environmental racism is mirrored in hundreds of struggles all over the world, from James Bay, Quebec to the Ecuadorian Amazon Rainforest. This new book documents these struggles and explores the underlying motivations and social forces that propel them. It concludes with a discussion of Native treaty rights and the next stage of the environmental movement.

250 pages, index
Paperback ISBN: 1-551640-00-7 $19.99
Hardcover ISBN: 1-551640-01-5 $48.99

WHEN FREEDOM WAS LOST
The Unemployed, the Agitator, and the State

Lorne Brown

This historical account of the little-known story of the jobless who drifted across the country during the Depression and were drawn into the work camps. Brown's factual and moving history records the desperation, disillusionment, and rebellion of these welfare inmates, and the repressive and shameful way in which they politicians and government authorities tried to keep the situation under control.

Lorne Brown seeks to remedy the dearth of the 30s labour Canadiana with this study of little-known labour camps.
Books in Canada

208 pages, photographs
Paperback ISBN: 0-920057-77-2 $14.99
Hardcover ISBN: 0-920057-75-6 $36.99

BETWEEN LABOR AND CAPITAL

Pat Walker, ed.

Essays on the general topic of class divisions in the U.S., giving both traditional and novel definitions.

337 pages
Paperback ISBN: 0-919618-86-3 $9.99
Hardcover ISBN: 0-919618-87-1 $19.99

FREE TRADE
Neither Free Nor About Trade

Christopher D. Merrett

Canada has been deeply marked by the Free Trade Agreement with the United States and has been brought increasingly under the sway of American trade policy, trade law, and multinational corporations.

The sweeping social and political changes that were initiated and accelerated in North American society as a result of the FTA are the subject of this book. Merrett looks at the mechanisms of Free Trade that are eroding the Nation-State and increasing regional and social disparities in Canada.

300 pages
Paperback ISBN: 1-551640-44-9   **$19.99**
Hardcover ISBN: 1-551640-45-7   **$48.99**

HOT MONEY AND THE POLITICS OF DEBT

R.T. Naylor

Introduction by Leonard Silk, former financial editor of the *New York Times*
2nd edition

Naylor discusses the global pool of hot and homeless money... how it is used and abused.
Journal of Economic Literature

As conspiracy theories go, here is one that is truly elegant. It involves everybody.
Washington Post

... a fascinating survey of international finance scams.
Globe and Mail

532 pages, index
Paperback ISBN: 1-895431-94-8   **$19.99**
Hardcover ISBN: 1-895431-95-6   **$48.99**

TOWARD A HUMANIST POLITICAL ECONOMY

Phillip Hansen and Harold Chorney

... the themes are relevant for those trying to fathom the post-Reaganite political world of the 1990s.
Canadian Book Review Annual

... their publication in one volume is a welcome addition to both the Canadian political economy literature and the literature on western Canada.
Prairie Forum

224 pages, index
Paperback ISBN:1-895431-22-0   **$19.99**
Hardcover ISBN:1-895431-23-9   **$48.99**